ADVANCED CALCULUS- LAPLACE TRANSFORMS

BY

EDWARD WALSH

WITH NUMEROUS SOLVED PROBLEMS COMPLETELY SOLVED IN DETAIL
WITH EVERY STEP INCLUDED

To actress Jean Harlow 1911-1937.

TABLE OF CONTNETS

Note for Librarians: A cataloguing record for this book is available from Library and Archives
Canada at www.collectionscanada.ca/amicus/index-e.html
ISBN 1-4120-8041-x

Offices in Canada, USA, Ireland and UK
This book was published *on-demand* in cooperation with Trafford Publishing. On-demand
publishing is a unique process and service of making a book available for retail sale to the
public taking advantage of on-demand manufacturing and Internet marketing. On-demand
publishing includes promotions, retail sales, manufacturing, order fulfilment, accounting and
collecting royalties on behalf of the author.

Book sales for North America and international:
Trafford Publishing, 6E–2333 Government St.,
Victoria, BC V8T 4P4 CANADA
phone 250 383 6864 (toll-free 1 888 232 4444)
fax 250 383 6804; email to orders@trafford.com
Book sales in Europe:
Trafford Publishing (UK) Limited, 9 Park End Street, 2nd Floor
Oxford, UK OXI IHH UNITED KINGDOM
phone 44 (0)1865 722 113 (local rate 0845 230 9601)
facsimile 44 (0)1865 722 868; info.uk@trafford.com
Order online at:
trafford.com/05-3039
10 9 8 7 6 5 4 3 2 1

Chapt 1 Introduction

The Laplace Transform And Its Application

To Solving Initial Value Problems

① the definition of the Laplace transformation of a function $f(x)$ is given by

$$L[f(x)] = \int_0^{+\infty} f(x) \, e^{-sx} \, dx$$

② from this definition we are able to develop a method for solving initial value problems involving linear differential equations with constant coefficients

③ in particular the Laplace transform may be used to solve initial value problems (I.V.P.'s) involving the following types of differential equations

1st order
⎡ ① first-order homogeneous linear D.E.'s with constant coefficients
⎣ ② first-order nonhomogeneous linear D.E.'s with constant coefficients

2nd order
⎡ ③ second-order homogeneous linear D.E.'s with constant coefficients
⎣ ④ second-order nonhomogeneous linear D.E.'s with constant coefficients

Chapt 1   Introduction

1st order homogeneous linear D.E with constant coefficients

$$a \frac{dy}{dx} + by = 0$$

$$y(0) = g \qquad \text{where } g \text{ is a number}$$

1st order non homogeneous linear D.E. with constant coefficients

$$a \frac{dy}{dx} + by = c(x)$$

$$y(0) = g \qquad \text{where } g \text{ is a number}$$

2nd order homogeneous linear D.E. with constant coefficients

$$a \frac{d^2 y}{dx^2} + b \frac{dy}{dx} + cy = 0$$

$$y(0) = g \qquad \text{where } g \text{ is a number}$$
$$y'(0) = h \qquad \text{where } h \text{ is a number}$$

2nd order non homogeneous linear D.E. with constant coefficients

$$a \frac{d^2 y}{dx^2} + b \frac{dy}{dx} + cy = d(x)$$

$$y(0) = g \qquad \text{where } g \text{ is a number}$$
$$y'(0) = h \qquad \text{where } h \text{ is a number}$$

Chapter Introduction

note

① ① the general solution of a first-order D.E.
contains one arbitrary constant

② the general solution of a second-order D.E.
contains two arbitrary constants

② ① an initial value problem involving
a first-order D.E. has one initial condition
$y(a) = b$ where a and b are numbers

② an initial value problem involving
a second-order D.E. has two initial conditions
$y(a) = b$ where a and b are numbers
$y'(a) = c$ where a and c are numbers

note

① the value of x in $y(a)$ and $y'(a)$ is the
same

② if the value of x is different the problem
is called a boundary value problem

③ Laplace transforms apply to initial value problems
not to boundary value problems

④ furthermore, when using the Laplace transform
to solve initial value problems we require
$a = 0$ in the above initial conditions
for first-order and second-order D.E.'s

chpt 1 Introduction
The Conventional Method For Solving
Initial Value Problems

① the conventional method to solve
first-order initial value problems (I.V.P.'s) is
① find the general solution of the D.E.
② apply the initial condition to determine
the particular value of the arbitrary constant
② the conventional method to solve
second-order initial value problems (I.V.P.'s) is
① find the general solution of the D.E.
② apply the initial conditions to determine
the particular values of the arbitrary constants

chpt 1   Introduction

The Advantage Of Using Laplace Transform
To Solve Initial Value Problems

① the advantage of using Laplace transform
to solve initial value problems is that
the general solution of the D.E. does not
have to be found

② the method produces the solution to the I.V.P.
with the proper values of the arbitrary constants
already in place

chapt 2 Functions And Superfunctions

Functions

① a function is a rule that assigns
to a given number a unique number

② the rule may be stated in words, however,
we usually use algebraic notation
to designate a function

③ for example the following algebraic expression
is a function

$$y = 3x - 2$$

the rule is to multiply the given number
by 3 and then subtract 2 from the product
if the given number is $x = 4$
then the unique number assigned to it is 10

$$y = 3x - 2$$
$$x = 4 \qquad \text{the given number}$$
$$y = 3(4) - 2$$
$$y = 12 - 2$$
$$y = 10 \qquad \text{the assigned unique number}$$

④ we may use arrow notation to symbolize this

$$4 \xrightarrow{b} 10$$

which states
the function f assigns to the number 4
a unique number 10

⑤ since $y = f(x)$ in function notation
we may substitute $f(x)$ for y
in the above discussion

(7)

chpt 2 Functions And Supervarctors

(6) the following algebraic expression
is a function
f(x) = 3x - 2
the rule again is to multiply the given number
by 3 and then subtract 2 from the product
if the given number is x = 4 then
the unique number assigned to it is 10

f(x) = 3x - 2
    x = 4

f(4) = 3·4 - 2
f(4) = 12 - 2
f(4) = 10

(7) using arrow notation we may write
4 $\xrightarrow{f}$ 10
which again states
the function f assigns to the number 4
the unique number 10

(8) note that with regard to a function
the input is a number   and
the output is also a number

(9) in the above examples
the input is 4
the output is 10

Chpt 2 Functions And Superfunctions

Superfunctions

① a superfunction is similar to a function
except
the input is a function and
the output is a function

② a superfunction is also called
ⓐ an operation   or
ⓑ a transformation

③ all three words -
superfunction
operation   and
transformation
mean the same thing

④ the latter two words are generally preferred

⑤ in general the letter T is used to indicate
that an operation or a transformation is to take place
on a function

⑥ T is called an operator

⑦ the definition of an operator is as follows -
an operator is a symbol placed before
a function to indicate an operation
is to be performed on the function
to produce a new function

⑧ using arrow notation we may write
$f(x) \overset{T}{\rightarrow} g(x)$

Chpt 2 Functions And Eigenfunctions

(9) In calculus there are two operations

① differentiation   and

② integration

(10) ① the operator associated with the operation of differentiation is given the symbol D

② the operator associated with the operation of integration is given the symbol I

operation	operator
differentiation	D
integration	I

(11) We will give an example of each operation with it's corresponding operator

(12) keep in mind
the input is a function   and
the output is a function

(13) also note the use of brackets [ ]
when applying the operator to a function

(14) the function is placed inside the brackets

& the operator is placed before the brackets

⑩

Unit 2 Functions And Exponential functions

ex differentiate the function $b(x) = 3x^5$

$$b(x) = 3x^5$$

$$D[b(x)] = D[3x^5]$$

$$= \frac{d}{dx}(3x^5)$$

$$= 15x^4$$

using arrow notation

$$3x^5 \xrightarrow{D} 15x^4$$

ex integrate the function $b(x) = 15x^4$

$$b(x) = 15x^4$$

$$I[b(x)] = I[15x^4]$$

$$= \int 15x^4 \, dx$$

$$= 15 \frac{x^5}{5} + C$$

$$= 3x^5 + C$$

let $C = 0$ (since C is arbitrary)

$$= 3x^5$$

using arrow notation

$$15x^4 \xrightarrow{I} 3x^5$$

Chapt 2 Functions And Superfunctions

⑮ we may summarize the preceding results

$$3X^5 \xrightarrow{D} 15X^4$$
$$15X^4 \xrightarrow[I]{} 3X^5$$

or

$$3X^5 \xrightarrow{D} 15X^4$$
$$3X^5 \xleftarrow[I]{} 15X^4$$

note

what one operator does the other operator undoes

D and I are referred to as inverse operators because of this property

~~Unit 2 Functions And Eigenfunctions~~

~~An Operator And It's Corresponding~~

~~Inverse Operator~~

① differentiation and integration are
inverse operations

therefore

D and I are inverse operators

(what one does the other undoes)

② we may write this property symbolically

$$f(x) \xrightarrow{D} g(x)$$
$$f(x) \xleftarrow{I} g(x)$$

also

$$f(x) \xrightarrow{I} g(x)$$
$$f(x) \xleftarrow{D} g(x)$$

Unit 3 The Laplace Transformation

The Laplace Transformation

① the definition of the Laplace transformation of the function $f(x)$ is given by

$$L[f(x)] = \int_0^{+\infty} f(x) e^{-px} dx$$

where

L is called the Laplace operator

② we require the function $f(x)$ to be defined for $x \geq 0$

③ the above is a definite integral in the variable x (technically an improper integral) and p is a constant

④ therefore
after evaluating the above integral for a particular function $f(x)$ we are left with a new function in p referred to by $F(p)$

⑤ note the capital F

⑥ using arrow notation we may write

$$f(x) \xrightarrow{L} F(p)$$

where

$f(x)$ is the input function and
$F(p)$ is the output function
L is the Laplace operator

⑦ it should be noted that the output function $F(p)$ is called the Laplace transform of the input function $f(x)$

⑧ the following example will illustrate finding the Laplace transform of the simplest function $f(x) = 1$

chapt 3 The Laplace Transformation

ex find the Laplace transform of the function

$\theta(x) = 1$

$$L[\theta(x)] = \int_0^{+\infty} \theta(x) \, e^{-px} \, dx$$

$\theta(x) = 1$

$$L[1] = \int_0^{+\infty} 1 \cdot e^{-px} \, dx$$

$$= \int_0^{+\infty} e^{-px} \, dx$$

$$= \left. \frac{1}{-p} e^{-px} \right|_0^{+\infty}$$

$$= \left. \left(- \frac{1}{p \, e^{px}} \right) \right|_0^{+\infty}$$

$$= \left(- \frac{1}{\infty} \right) - \left(- \frac{1}{p \, e^0} \right)$$

$$= (-0) - \left[- \frac{1}{p(1)} \right]$$

$$= 0 + \frac{1}{p}$$

$$= \frac{1}{p}$$

note

$+\infty = \infty$

(15)

Unit 3 The Laplace Transformation using substitution

$$L[b(x)] = \int_0^{+\infty} b(x) \, e^{-px} \, dx$$

$$L[1] = \int_0^{+\infty} 1 \cdot e^{-px} \, dx$$

$$= \int_0^{+\infty} e^{-px} \, dx$$

$$\int e^{-px} \, dx$$

$$\text{let } u = -px$$
$$du = -p \, dx$$
$$dx = \frac{1}{-p} \, du$$

$$\int e^u \left(\frac{1}{-p} \, du \right)$$

$$= -\frac{1}{p} \int e^u \, du$$

$$= -\frac{1}{p} (e^u + c)$$

$$= -\frac{1}{p} e^u + C$$

$$= -\frac{1}{p} e^{-px} + C$$

Unit 3 The Laplace Transformation ⑯ $\quad$ Intro

$$\int_0^{+\infty} e^{-\rho x}\, dx$$

$$= \left(-\frac{1}{\rho}\, e^{-\rho x} \right) \Big|_0^{+\infty}$$

$$= \left(-\frac{1}{\rho e^{\rho x}} \right) \Big|_0^{+\infty}$$

$$= \left(-\frac{1}{\infty} \right) - \left(-\frac{1}{\rho e^{0}} \right)$$

$$(-0) - \left[-\frac{1}{\rho(1)} \right]$$

$$= \quad 0 \;+\; \frac{1}{\rho}$$

$$= \quad \frac{1}{\rho}$$

Chapt 3 The Laplace Transformation

note

① the input function is $f(x) = 1$
the output function is $F(p) = \dfrac{1}{p}$

② using arrow notation we may write

$f(x) \xrightarrow{L} F(p)$

$1 \xrightarrow{L} \dfrac{1}{p}$

where
the Laplace operator L transforms
the function $f(x) = 1$ into the function $F(p) = \dfrac{1}{p}$

Chapt 3 The Laplace Transformation

ex find the Laplace transform of the function
$f(x) = K$ where K is a constant

$$L[f(x)] = \int_0^{+\infty} f(x) \, e^{-px} \, dx$$

$$f(x) = K$$

$$L[K] = \int_0^{+\infty} K \cdot e^{-px} \, dx$$

$$= K \left(\frac{1}{-p} \right) e^{-px} \Big|_0^{+\infty}$$

$$= \left(-\frac{K}{p \, e^{px}} \right) \Big|_0^{+\infty}$$

$$= \left(-\frac{K}{\infty} \right) - \left(-\frac{K}{p \, e^0} \right)$$

$$= (-0) - \left[-\frac{K}{p(1)} \right]$$

$$= 0 + \frac{K}{p}$$

$$= \frac{K}{p}$$

Chapt 3 The Laplace Transformation
using substitution

$$L\left[f(x)\right] = \int_0^{+\infty} f(x)\, e^{-px}\, dx$$

$$L\left[K\right] = \int_0^{+\infty} K \cdot e^{-px}\, dx$$

$$= K \int \frac{\int K\, e^{-px}\, dx}{e^{-px}\, dx}$$

let $u = -px$

$$du = -p\, dx$$

$$dx = \frac{1}{-p}\, du$$

$$K \int e^{u}\left(\frac{1}{-p}\, du\right)$$

$$= -\frac{K}{p} \int e^{u}\, du$$

$$= -\frac{K}{p}\left(e^{u} + c\right)$$

$$= -\frac{K}{p}\, e^{u} + c$$

$$= -\frac{K}{p}\, e^{-px} + c$$

Chapt 3 The Laplace Transform (20)

$$\int_0^{+\infty} K e^{-px} \, dx$$

$$= \left(-\frac{K}{p} e^{-px} \right) \Big|_0^{+\infty}$$

$$= \left(-\frac{K}{pe^{px}} \right) \Big|_0^{+\infty}$$

$$= \left(-\frac{K}{\infty} \right) - \left(-\frac{K}{pe^0} \right)$$

$$= (-0) - \left[-\frac{K}{p(1)} \right]$$

$$= 0 + \frac{K}{p}$$

$$= \frac{K}{p}$$

Chapt 3 The Laplace Transformation

note:

(1) ① the input function → $b(x) = K$

② the output function → $F(p) = \dfrac{K}{p}$

② using arrow notation we may write

$b(x) \xrightarrow{L} F(p)$

$K \xrightarrow{L} \dfrac{K}{p}$

where

$L \to$ the Laplace operation

Unit 3 The Laplace Transform

The Laplace Transform Of zero

$$L[f(x)] = \int_0^{+\infty} f(x)\, e^{-px}\, dx$$

$$L[0] = \int_0^{+\infty} 0 \cdot e^{-px}\, dx$$

$$= 0 \int_0^{+\infty} e^{-px}\, dx$$

$$= 0 \left[\left(-\frac{1}{p} e^{-px} \right) \Big|_0^{+\infty} \right]$$

$$= 0 \left[\left(-\frac{1}{p\, e^{px}} \right) \Big|_0^{+\infty} \right]$$

$$= 0 \left(-\frac{1}{+\infty} + \frac{1}{p\, e^0} \right)$$

$$= 0 \left(-0 + \frac{1}{p} \right)$$

$$= 0 \left(\frac{1}{p} \right)$$

$$= 0$$

also

$$L[K] = \frac{K}{p}$$

$$L[0] = \frac{0}{p} = 0$$

Chapt 3 The Laplace Transformation

In Summary we have derived the following two formulas

$$L[1] = \frac{1}{p}$$

$$L[K] = \frac{K}{p} \qquad \text{where } K \text{ is a constant}$$

also
we have established

$$L[0] = 0$$

The Laplace Transformation Of

Chapt 4 Bore Elementary Function

The Laplace Transformation Of

Bore Elementary Functions

① we shall use the definition of
the Laplace transformation of the function $f(x)$

$$L\left[f(x)\right] = \int_0^{+\infty} f(x)\, e^{-px}\, dx$$

to find the Laplace transform of each
of the following elementary functions

polynomial functions

$f(x) = 1$

$f(x) = x$

$f(x) = x^2$

$f(x) = x^3$

$f(x) = x^4$

$\vdots$

$f(x) = x^n$ where $n = 0, 1, 2, 3, \ldots$

exponential functions

$f(x) = e^x$

$f(x) = e^{kx}$ where k is a constant

trigonometric functions

$f(x) = \sin x$

$f(x) = \cos x$

$f(x) = \sin kx$ where k is a constant

$f(x) = \cos kx$

(25)

The Laplace Transformation of &Intro

Unit 4 Basic Elementary Functions

hyperbolic functions

$f(x) = \sinh x$

$f(x) = \cosh x$

$f(x) = \sinh kx$

$f(x) = \cosh kx$ where k is a constant

The Laplace Transformation Of

Chapt 5 $f(x) = x^n$ where $n = 0, 1, 2, 3, \ldots$

The Laplace Transformation Of The Function

$f(x) = x^n$ where $n = 0, 1, 2, 3, \ldots$

① In this chapter we shall use the definition of the Laplace transformation

$$L[f(x)] = \int_0^{+\infty} f(x)\, e^{-px}\, dx$$

to find the Laplace transforms of each of the following functions

$f(x) = 1$

$f(x) = x$

$f(x) = x^2$

$f(x) = x^3$

$f(x) = x^4$

$\vdots$

$f(x) = x^n$ where $n = 0, 1, 2, 3, \ldots$

② our ultimate goal is to derive the formula

$$L[x^n] = \frac{n!}{p^{n+1}} \qquad n = 0, 1, 2, 3, \ldots$$

The Laplace Transformation Of

Chapts 5 $f(x) = x^n$ $n = 0, 1, 2, 3, \ldots$

① Find the Laplace transform of the function
$f(x) = 1$

$$L[f(x)] = \int_0^{+\infty} f(x) \, e^{-px} \, dx$$

<u>$f(x) = 1$</u>

$$L[1] = \int_0^{+\infty} 1 \cdot e^{-px} \, dx$$

$$= \int_0^{+\infty} e^{-px} \, dx$$

$$= \left(-\frac{1}{p} e^{-px} \right) \Big|_0^{+\infty}$$

$$= \left(-\frac{1}{p \, e^{px}} \right) \Big|_0^{+\infty}$$

$$= \left(-\frac{1}{\infty} \right) - \left(-\frac{1}{p \, e^0} \right)$$

$$= (-0) - \left[-\frac{1}{p(1)} \right]$$

$$= 0 + \frac{1}{p}$$

$$= \frac{1}{p}$$

$$\underline{\text{The Laplace Transformation of}}$$
$$\underline{\text{Chapt 5} \qquad f(x) = x^n \qquad n = 0, 1, 2, 3, \ldots}$$

② find the Laplace transform of the function

$$f(x) = x$$

$$L[f(x)] = \int_0^{+\infty} f(x)\, e^{-px}\, dx$$

$\underline{f(x) = x}$

$$L[x] = \int_0^{+\infty} x\, e^{-px}\, dx$$

The Laplace Transform of
$$f(x) = x^n \qquad n = 0, 1, 2, 3, \ldots$$

Chapt 5

(26.1)

$$\int x \, e^{-px} \, dx$$

$$\int u \, dv = uv - \int v \, du$$

let $u = x$ $\qquad\qquad dv = e^{-px} \, dx$

$\qquad\qquad du = dx$ $\qquad\qquad \int dv = \int e^{-px} \, dx$

$f(x) = x$ $\qquad\qquad\qquad\qquad v = -\dfrac{1}{p} e^{-px}$

$$\int x \, e^{-px} \, dx$$

$$= x \left(-\frac{1}{p} e^{-px} \right) - \int \left(-\frac{1}{p} e^{-px} \right) dx$$

$$= -\frac{x}{p \, e^{px}} + \frac{1}{p} \int e^{-px} \, dx$$

II C

chpt 5 $\dfrac{\text{The Laplace Transform of}}{f(x) = x^n \quad n = 0, 1, 2, 3, \ldots}$

2 cont $\displaystyle\int_0^{+\infty} x \, e^{-px} \, dx$

$$= \left(-\frac{x}{p\,e^{px}} \right)\Big|_0^{+\infty} + \frac{1}{p} \int_0^{+\infty} e^{-px} \, dx$$

$f(x) = x \quad = \left(-\dfrac{x}{p\,e^{px}} \right)\Big|_0^{+\infty} + \dfrac{1}{p}\left(\dfrac{1}{p} \right)$

$$= \left(-\frac{x}{p\,e^{px}} \right)\Big|_0^{+\infty} + \frac{1}{p^2}$$

$$= -\left(\frac{x}{p\,e^{px}}\Big|_0^{+\infty} \right) + \frac{1}{p^2}$$

II D

The Laplace Transformation $\%$

Cont 5 $\phi(x) = x^n$ $n = 0, 1, 2, 3, \cdots$

$\textcircled{2 cont}$ $\left. \dfrac{x}{p e^{px}} \right|_0^{+\infty}$

$= \dfrac{\infty}{\infty} - \dfrac{0}{p e^0}$

$\phi(x) = x \quad = \dfrac{\infty}{\infty} - \dfrac{0}{p (1)}$

$= \dfrac{\infty}{\infty} - \dfrac{0}{p}$

$= \dfrac{\infty}{\infty} - 0$

$= \dfrac{\infty}{\infty}$

$= \lim_{x \to +\infty} \dfrac{x}{p e^{px}}$ use L'Hospitals Law

$= \lim_{x \to +\infty} \dfrac{1}{p e^{px} \cdot p}$

$= \lim_{x \to +\infty} \dfrac{1}{p^2 e^{px}}$

$= 0$

II E

Prob

$\underline{\text{ch. }t5}$ The Laplace Transformation of
$f(x) = x^n \qquad n = 0, 1, 2, 3, \ldots$

(2 cont) $- \quad 0 \quad + \quad \dfrac{1}{p^2}$

$= \quad \dfrac{1}{p^2}$

$f(x) = x$

The Laplace Transformation of

Chapt 5 $f(x) = x^n$ $n = 0, 1, 2, 3, \ldots$

③ find the Laplace transform of the function
$f(x) = x^2$

$f(x) = x^2$

$$L\left[f(x)\right] = \int_0^{+\infty} f(x)\, e^{-px}\, dx$$

$$L\left[x^2\right] = \int_0^{+\infty} x^2\, e^{-px}\, dx$$

$$\underline{\text{The Laplace Transformation of}}$$

$$\underline{\text{Chapt 5} \quad f(x) = x^n \quad \quad n = 0, 1, 2, 3, \dots}$$

(3 cont) $\qquad \int x^2 e^{-px} \, dx$

$$\int u \, dv = uv - \int v \, du$$

$$\text{let } u = x^2 \qquad \qquad dv = e^{-px} \, dx$$

$f(x) = x^2$

$$du = 2x \, dx \qquad \qquad \int dv = \int e^{-px} \, dx$$

$$v = -\frac{1}{p} e^{-px}$$

$$\int x^2 e^{-px} \, dx$$

$$= x^2 \left(-\frac{1}{p} e^{-px} \right) - \int \left(-\frac{1}{p} e^{-px} \right) 2x \, dx$$

$$= -\frac{x^2}{p \, e^{px}} + \frac{2}{p} \int x \, e^{-px} \, dx$$

III C

35

Prob V

$$\underline{\text{chapt 5}} \quad \underline{\text{The Laplace Transform of}} \\ \underline{\quad f(x) = x^n \quad n = 0, 1, 2, 3, \ldots}$$

③ cont
$$\int_0^{+\infty} x^2 e^{-px} \, dx$$

$$= \left(-\frac{x^2}{p e^{px}} \right) \Big|_0^{+\infty} + \frac{2}{p} \int_0^{+\infty} x e^{-px} \, dx$$

$$f(x) = x^2 = \left(-\frac{x^2}{p e^{px}} \right) \Big|_0^{+\infty} + \frac{2}{p} \left(\frac{1}{p^2} \right)$$

$$= \left(-\frac{x^2}{p e^{px}} \right) \Big|_0^{+\infty} + \frac{2}{p^3}$$

$$= -\left(\frac{x^2}{p e^{px}} \Big|_0^{+\infty} \right) + \frac{2}{p^3}$$

Con't $\quad \underline{\text{The Laplace Transformation Of}}$

③⑥

$$\underline{G(x) = x^n \quad n = 0, 1, 2, 3, \dots}$$

③ Con't

$$\frac{x^2}{p\,e^{px}}\bigg|_0^{+\infty}$$

$$= \quad \frac{\infty}{\infty} - \frac{0}{p\,e^0}$$

$G(x) = x^2 \quad = \quad \dfrac{\infty}{\infty} - \dfrac{0}{p(1)}$

$$= \quad \frac{\infty}{\infty} - 0$$

$$= \quad \frac{\infty}{\infty}$$

$$= \quad \lim_{x \to +\infty} \frac{x^2}{p\,e^{px}} \qquad\qquad \text{use L'Hospitals Law}$$

$$= \quad \lim_{x \to +\infty} \frac{2x}{p\,e^{px} \cdot p}$$

$$= \quad \lim_{x \to +\infty} \frac{2x}{p^2\,e^{px}}$$

$$= \quad \lim_{x \to +\infty} \frac{2}{p^2\,e^{px} \cdot p}$$

$$= \quad \lim_{x \to +\infty} \frac{2}{p^3\,e^{px}}$$

$$= \quad \frac{2}{\infty}$$

$$= \quad 0$$

Unit 5

The Laplace Transformation Of

$$f(x) = x^n \qquad n = 0, 1, 2, 3, \ldots$$

(3 cont)

$$- \quad 0 \quad + \quad \frac{2}{p^3}$$

$$= \quad \frac{2}{p^3}$$

$$f(x) = x^2$$

The Laplace Transformation of
chpt 5 $f(x) = x^n$ $n = 0, 1, 2, 3, \ldots$

4) find the Laplace transform of the function
$f(x) = x^3$

$f(x) = x^3$

$$L[f(x)] = \int_0^{+\infty} f(x) \, e^{-px} \, dx$$

$$L[x^3] = \int_0^{+\infty} x^3 \, e^{-px} \, dx$$

Chapt 5   The Laplace Transformation of

$f(x) = x^n$ $n = 0, 1, 2, 3, \ldots$

(4 cont)

$$\int x^3 e^{-px} \, dx$$

$f(x) = x^3$

$$\int u \, dv = uv - \int v \, du$$

let $u = x^3$ $dv = e^{-px} \, dx$

$du = 3x^2 \, dx$ $\int dv = \int e^{-px} \, dx$

$$v = -\frac{1}{p} e^{-px}$$

$$\int x^3 e^{-px} \, dx$$

$$= x^3 \left(-\frac{1}{p} e^{-px} \right) - \int \left(-\frac{1}{p} e^{-px} \right) 3x^2 \, dx$$

$$= -\frac{x^3}{p \, e^{px}} + \frac{3}{p} \int x^2 e^{-px} \, dx$$

Chapt 5 $\qquad$ The Laplace Transformation Of ⑩ Prob

$$f(x) = x^n \qquad n = 0, 1, 2, 3, \ldots$$

④ cont

$$\int_0^{+\infty} x^3 \, e^{-px} \, dx$$

$f(x) = x^3$

$$= \left(-\frac{x^3}{p\,e^{px}} \right)\Big|_0^{+\infty} + \frac{3}{p} \int_0^{+\infty} x^2 \, e^{-px} \, dx$$

$$= \left(-\frac{x^3}{p\,e^{px}} \right)\Big|_0^{+\infty} + \frac{3}{p} \left(\frac{2}{p^3} \right)$$

$$= \left(-\frac{x^3}{p\,e^{px}} \right)\Big|_0^{+\infty} + \frac{6}{p^4}$$

$$= -\left(\frac{x^3}{p\,e^{px}} \Big|_0^{+\infty} \right) + \frac{6}{p^4}$$

Chapt 5

(41)

Prob

The Laplace Transformation Of
$f(x) = x^n$ $n = 0, 1, 2, 3, \ldots$

(4 cont)

$$\left. \frac{x^3}{p \, e^{px}} \right|_0^{+\infty}$$

$$\frac{\infty}{\infty} - \frac{0}{p e^0}$$

$f(x) = x^3$

$$\frac{\infty}{\infty} - \frac{0}{p(1)}$$

$$\frac{\infty}{\infty} - 0$$

$$\frac{\infty}{\infty}$$

$$\lim_{x \to +\infty} \frac{x^3}{p \, e^{px}}$$ use L'Hospital's Law

$$\lim_{x \to +\infty} \frac{3x^2}{p^2 \, e^{px}}$$

$$\lim_{x \to +\infty} \frac{6x}{p^3 \, e^{px}}$$

$$\lim_{x \to +\infty} \frac{6}{p^4 \, e^{px}}$$

$$\frac{6}{\infty}$$

$$0$$

IV E

const

The Laplace Transformation of
$f(x) = x^n$ $n = 0, 1, 2, 3, ...$

$\boxed{4cot}$ — 0 + $\dfrac{6}{p^4}$

$= \dfrac{6}{p^4}$

$f(x) = x^3$

V A

The Laplace Transformation of

Chapt 5 $f(x) = x^n$ $n = 0, 1, 2, 3, \ldots$

⑤ find the Laplace transform of the function

$f(x) = x^4$

$f(x) = x^4$

$$L[f(x)] = \int_0^{+\infty} f(x)\, e^{-px}\, dx$$

$$L[x^4] = \int_0^{+\infty} x^4\, e^{-px}\, dx$$

(44)

Ch-pt 5   The Laplace Transform of

$\theta(x) = x^n$ $n = 0, 1, 2, 3, \ldots$

(5 cont) $\int x^4 e^{-px} \, dx$

$\theta(x) = x^4$

$$\int u \, dv = uv - \int v \, du$$

$$\text{let } u = x^4 \qquad\qquad dv = e^{-px} \, dx$$

$$du = 4x^3 \, dx \qquad \int dv = \int e^{-px} \, dx$$

$$v = -\frac{1}{p} e^{-px}$$

$$\int x^4 e^{-px} \, dx$$

$$= x^4 \left(-\frac{1}{p} e^{-px} \right) - \int \left(-\frac{1}{p} e^{-px} \right) 4x^3 \, dx$$

$$= -\frac{x^4}{p e^{px}} + \frac{4}{p} \int x^3 e^{-px} \, dx$$

Chapt 5   The Laplace Transformation of

$$b(x) = x^n \qquad n = 0, 1, 2, 3, \dots$$

(cont)

$$\int_0^{+\infty} x^4 e^{-px} \, dx$$

$$= \left(-\frac{x^4}{p\, e^{px}} \right)\Big|_0^{+\infty} + \frac{4}{p} \int_0^{+\infty} x^3 e^{-px} \, dx$$

$b(x) = x^4 =$
$$\left(-\frac{x^4}{p\, e^{px}} \right)\Big|_0^{+\infty} + \frac{4}{p} \left(\frac{6}{p^4} \right)$$

$$= \left(-\frac{x^4}{p\, e^{px}} \right)\Big|_0^{+\infty} + \frac{24}{p^5}$$

$$= -\left(\frac{x^4}{p\, e^{px}} \Big|_0^{+\infty} \right) + \frac{24}{p^5}$$

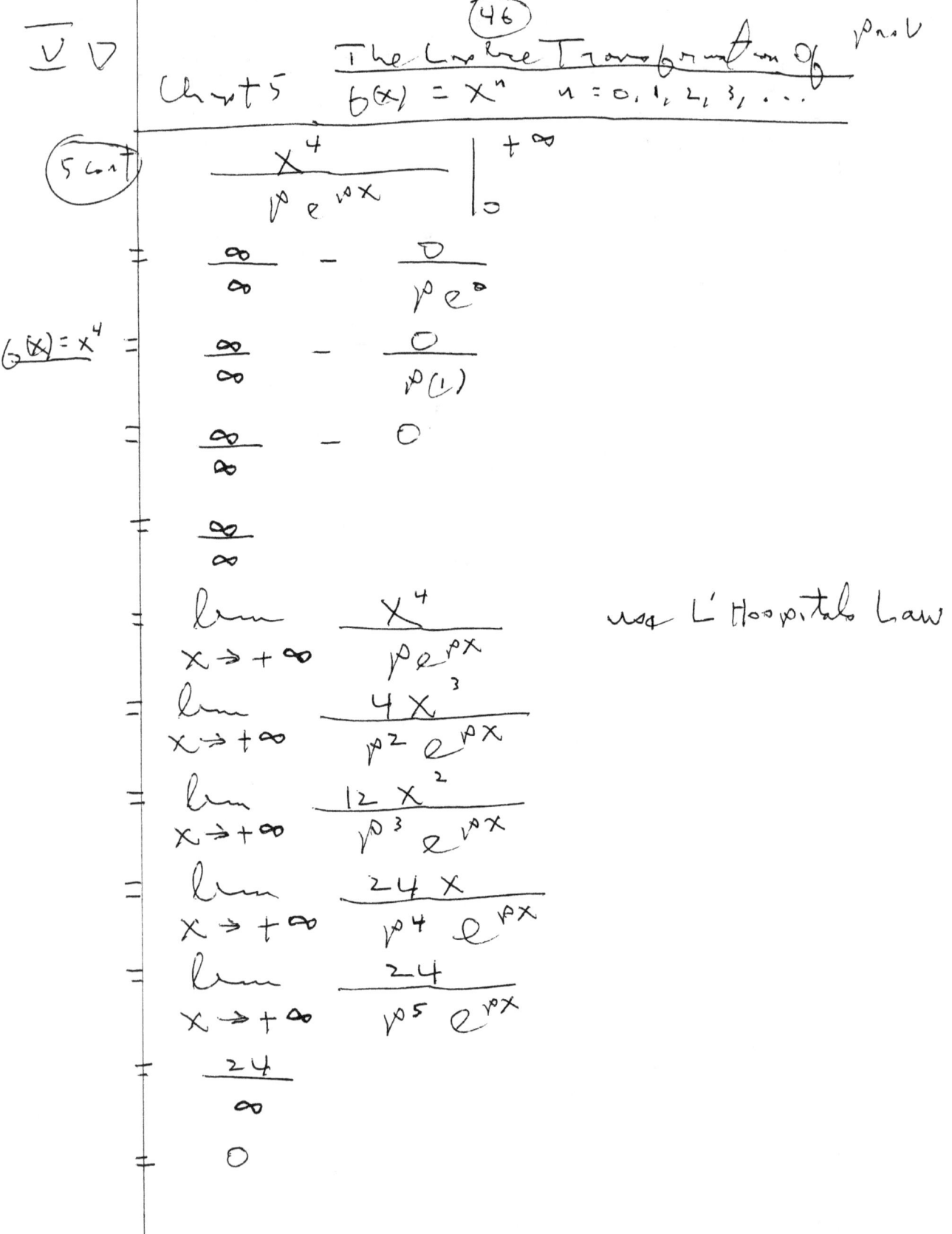

$\overline{\upsilon}\,\upsilon$ | Chapt 5   The Laplace Transformation of ϕnal υ

$$b(x) = x^n \qquad n = 0, 1, 2, 3, \ldots$$

(Scant)

$$\left. \frac{x^4}{\rho e^{\rho x}} \right|_0^{+\infty}$$

$$= \frac{\infty}{\infty} - \frac{0}{\rho e^0}$$

$b(x) = x^4 \;=\; \dfrac{\infty}{\infty} - \dfrac{0}{\rho(1)}$

$$= \frac{\infty}{\infty} - 0$$

$$= \frac{\infty}{\infty}$$

$$= \lim_{x \to +\infty} \frac{x^4}{\rho e^{\rho x}} \qquad\qquad \text{use L'Hospital's Law}$$

$$= \lim_{x \to +\infty} \frac{4x^3}{\rho^2 e^{\rho x}}$$

$$= \lim_{x \to +\infty} \frac{12x^2}{\rho^3 e^{\rho x}}$$

$$= \lim_{x \to +\infty} \frac{24x}{\rho^4 e^{\rho x}}$$

$$= \lim_{x \to +\infty} \frac{24}{\rho^5 e^{\rho x}}$$

$$= \frac{24}{\infty}$$

$$= 0$$

<u>J E</u>

Chapts $\quad$ <u>The Laplace Transform of</u> $\quad$ Prob ⑰

$$\frac{G(x) = x^n}{\qquad\qquad\qquad n = 0, 1, 2, 3, \ldots}$$

(5cont) $\quad - 0 + \dfrac{24}{p^5}$

$\qquad = \dfrac{24}{p^5}$

$G(x) = x^4$

Prob

The Laplace Transformation of

Chapt 5 $f(x) = x^n$ $n = 0, 1, 2, 3, \ldots$

⑥

The Pattern for $f(x) = x^n$ $n = 0, 1, 2, 3, \ldots$

$L[1] = L[x^0] = \dfrac{1}{p} = \dfrac{0!}{p^{0+1}}$

$f(x) = x^n$

$L[x] = \dfrac{1}{p^2} = \dfrac{1!}{p^{1+1}}$

$L[x^2] = \dfrac{2}{p^3} = \dfrac{2!}{p^{2+1}}$

$L[x^3] = \dfrac{6}{p^4} = \dfrac{3!}{p^{3+1}}$

$L[x^4] = \dfrac{24}{p^5} = \dfrac{4!}{p^{4+1}}$

$\vdots$

$L[x^n] = \dfrac{n!}{p^{n+1}}$

The Laplace Transformation Of Chapt 6 Exponential Functions

The Laplace Transformation Of Exponential Functions

① In this chapter we shall use the definition of the Laplace Transformation

$$L[f(x)] = \int_0^{+\infty} f(x)\, e^{-px}\, dx$$

to find the Laplace transforms of the following exponential functions

$$f(x) = e^x$$
$$f(x) = e^{kx}$$

② In particular we want to derive the formulas

$$L[e^x] = \frac{1}{p-1}$$

$$L[e^{kx}] = \frac{1}{p-k}$$

I - A

The Laplace Transformation of
Chapt6 Exponential Functions

① Use the definition of the Laplace transformation
to find the Laplace transform of
$f(x) = e^x$

$f(x) = e^x$

$$L[f(x)] = \int_0^{+\infty} f(x) \, e^{-px} \, dx$$

$$L[e^x] = \int_0^{+\infty} e^x \cdot e^{-px} \, dx$$

$$= \int_0^{+\infty} e^{x-px} \, dx$$

$$= \int_0^{+\infty} e^{(1-p)x} \, dx$$

$$= \int_0^{+\infty} e^{-(p-1)x} \, dx$$

$$= \left. - \left(\frac{1}{p-1} \right) e^{-(p-1)x} \right|_0^{+\infty}$$

$$= \left. - \frac{1}{(p-1) \, e^{(p-1)x}} \right|_0^{+\infty}$$

$$= - \frac{1}{\infty} + \frac{1}{(p-1) \, e^0}$$

$$= - 0 + \frac{1}{(p-1)(1)}$$

$$= \frac{1}{p-1}$$

I B

Proof

The Laplace Transformation of
Exponential Functions
using substitution

$g(x) = e^x$

$$L[g(x)] = \int_0^{+\infty} g(x)\, e^{-px}\, dx$$

$$L[e^x] = \int_0^{+\infty} e^x \cdot e^{-px}\, dx$$

$$= \int_0^{+\infty} e^{x-px}\, dx$$

$$= \int_0^{+\infty} e^{(1-p)x}\, dx$$

$$= \int_0^{+\infty} e^{-(p-1)x}\, dx$$

IC

Prob

(52)

Chpt 6 The Laplace Transformation of Exponential Functions

(1 cont)

$$\int e^{-(p-1)x}\, dx$$

$$\text{let } u = -(p-1)x$$
$$du = -(p-1)\, dx$$
$$dx = -\left(\frac{1}{p-1}\right) du$$

$b(x) = e^{x}$

$$\int e^{u}\left[-\left(\frac{1}{p-1}\right) du\right]$$

$$= -\left(\frac{1}{p-1}\right) \int e^{u}\, du$$

$$= -\left(\frac{1}{p-1}\right)(e^{u} + c)$$

$$= -\left(\frac{1}{p-1}\right) e^{u} + c$$

$$= -\left(\frac{1}{p-1}\right) e^{-(p-1)x} + c$$

The Laplace Transformation of Exponential Function

Cont.

1 cont.

$$- \left(\frac{1}{p-1} \right) e^{-(p-1)x} \Big|_0^{+\infty}$$

$$= - \frac{1}{(p-1) e^{(p-1)x}} \Big|_0^{+\infty}$$

$b(x) = e^x =$

$$- \frac{1}{\infty} + \frac{1}{(p-1) e^0}$$

$$= - 0 + \frac{1}{(p-1)(1)}$$

$$= \frac{1}{p-1}$$

The Laplace Transformation of

chpt 6   Exponential Functions

(1 cont)

note

we may write

$$f(x) \xrightarrow{L} F(p)$$

$f(x) = e^x$

$$e^x \xrightarrow{L} \frac{1}{p-1}$$

where

$$f(x) = e^x \qquad \text{and}$$

$$F(p) = \frac{1}{p-1}$$

The Laplace Transformation of Exponential Functions

② use the definition of the Laplace transformation to find the Laplace transform of
$$\phi(x) = e^{kx}$$ where k is a constant

$\phi(x) = e^{kx}$

$$L[\phi(x)] = \int_0^{+\infty} \phi(x)\, e^{-px}\, dx$$

$$L[e^{kx}] = \int_0^{+\infty} e^{kx} \cdot e^{-px}\, dx$$

$$= \int_0^{+\infty} e^{kx - px}\, dx$$

$$= \int_0^{+\infty} e^{kx - px}\, dx$$

$$= \int_0^{+\infty} e^{(k-p)x}\, dx$$

$$= \int_0^{+\infty} e^{-(p-k)x}\, dx$$

$$= -\left(\frac{1}{p-k}\right) e^{-(p-k)x} \Big|_0^{+\infty}$$

$$= -\frac{1}{(p-k)\, e^{(p-k)x}} \Big|_0^{+\infty}$$

$$= -\frac{1}{\infty} + \frac{1}{(p-k)\, e^0}$$

$$= -0 + \frac{1}{(p-k)(1)}$$

$$= \frac{1}{p-k}$$

(2cont)

(56)

The Laplace Transformation of Chapt 6 Exponential Functions using substitution

$$L[f(x)] = \int_0^{+\infty} f(x) \, e^{-px} \, dx$$

$f(x) = e^{kx}$

$$L[e^{kx}] = \int_0^{+\infty} e^{kx} \cdot e^{-px} \, dx$$

$$= \int_0^{+\infty} e^{kx - px} \, dx$$

$$= \int_0^{+\infty} e^{(k-p)x} \, dx$$

$$= \int_0^{+\infty} e^{-(p-k)x} \, dx$$

Chpt 6

The Laplace Transformation of
Exponential Functions

$$\int e^{-(p-k)x} \, dx$$

$$\text{let} \quad u = -(p-k)x$$

$$du = -(p-k)\, dx$$

$$dx = -\left(\frac{1}{p-k}\right) du$$

$$\int e^{u} \left[-\left(\frac{1}{p-k}\right)\right] du$$

$$\int e^{u} \, du$$

$$= -\left(\frac{1}{p-k}\right)$$

$$= -\left(\frac{1}{p-k}\right)(e^{u}+C)$$

$$= -\left(\frac{1}{p-k}\right)e^{u} + C$$

$$= -\left(\frac{1}{p-k}\right)e^{-(p-k)x} + C$$

$$\overbrace{58}$$

The Laplace Transformation of Exponential Functions

Cont'd

$\boxed{2 \text{ cont}}$

$$- \left(\frac{1}{p-k} \right) e^{-(p-k)x} \Big|_0^{+\infty}$$

$$= - \frac{1}{(p-k) e^{(p-k)x}} \Big|_0^{+\infty}$$

$b(x) = e^{kx}$

$$= - \frac{1}{\infty} + \frac{1}{(p-k) e^0}$$

$$= -0 + \frac{1}{(p-k)(1)}$$

$$= \frac{1}{p-k}$$

II E

Prob

The Laplace Transformation of Unto Exponential Functions

(2 cont) note

we may write

$$f(x) \xrightarrow{L} F(\rho)$$

$f(x) = e^{Kx}$

$$e^{Kx} \xrightarrow{L} \frac{1}{\rho - K}$$

where

$$g(x) = e^{Kx}$$

$$F(\rho) = \frac{1}{\rho - K}$$

60
The Laplace Transformation of
Chapt 6 Exponential Functions

In summary we have derived the formulas

$$L[e^x] = \frac{1}{p-1}$$

$$L[e^{Kx}] = \frac{1}{p-K}$$
where K is a constant

which are the Laplace transforms of e^x and e^{Kx} respectively

The Laplace Transformation Of Trigonometric Functions

The Laplace Transformation Of Trigonometric Functions

① In this chapter we shall use the definition of the Laplace transformation

$$L[f(x)] = \int_0^{+\infty} f(x)\, e^{-px}\, dx$$

to derive the Laplace transform of each of the following trigonometric functions

$$f(x) = \sin x$$
$$f(x) = \cos x$$

$$f(x) = \sin Kx \qquad \text{where } K \text{ is a constant}$$
$$f(x) = \cos Kx \qquad \text{where } K \text{ is a constant}$$

The Laplace Transformation 0/6
chapt 7 Trigonometric Functions

② the results we want to obtain one

$f(x) = \sin x$

$L[\sin x] = \dfrac{1}{p^2 + 1}$

$f(x) = \cos x$

$L[\cos x] = \dfrac{p}{p^2 + 1}$

$f(x) = \sin kx$ where k is a constant

$L[\sin kx] = \dfrac{k}{p^2 + k^2}$

$f(x) = \cos kx$ where k is a constant

$L[\cos kx] = \dfrac{p}{p^2 + k^2}$

I A

The Laplace Transformation of

Chapt 7 Trigonometric Function

① use the definition of the Laplace transformation
to find the Laplace transform of

$$f(x) = \sin x$$

$f(x) =$
$\sin x$

$$L[f(x)] = \int_0^{+\infty} f(x) \, e^{-\nu x} \, dx$$

$$L[\sin x] = \int_0^{+\infty} \sin x \, e^{-\nu x} \, dx$$

I B

The Laplace Transformation of

Chapt 7 Trigonometric Functions

$$\int \sin x \; e^{-px} \, dx$$

(cont)

$$\int u \, dv = uv - \int v \, du$$

$$\text{let } u = \sin x \qquad dv = e^{-px} \, dx$$

$$du = \cos x \, dx \qquad \int dv = \int e^{-px} \, dx$$

$$v = -\frac{1}{p} e^{-px}$$

$6(x) =$
$\sin x$

$$\int \sin x \; e^{-px} \, dx$$

$$= \sin x \left(-\frac{1}{p} e^{-px} \right) - \int \left(-\frac{1}{p} e^{-px} \right) \cos x \, dx$$

$$= -\frac{\sin x}{p \, e^{px}} + \frac{1}{p} \int \cos x \; e^{-px} \, dx$$

IC

(69)

Prob

The Laplace Transformation Of

Chapt 7 Trigonometric Functions

(cont)

$$\int \cos x \, e^{-px} \, dx$$

$$\int u \, dv = uv - \int v \, du$$

$f(x) =$

$\sin x$

$$\text{let } u = \cos x \qquad dv = e^{-px} \, dx$$

$$du = -\sin x \, dx \qquad \int dv = \int e^{-px} \, dx$$

$$v = -\frac{1}{p} e^{-px}$$

$$\int \cos x \, e^{-px} \, dx$$

$$= \cos x \left(-\frac{1}{p} e^{-px} \right) - \int \left(-\frac{1}{p} e^{-px} \right) (-\sin x \, dx)$$

$$= \frac{\cos x}{p \, e^{px}} - \frac{1}{p} \int \sin x \, e^{-px} \, dx$$

IV

(66)

<u>The Laplace Transformation of</u>
Chpt 7 <u>Trigonometric Functions</u>

(1 cont) $\int \sin x \, e^{-px} \, dx$

$= -\dfrac{\sin x}{p \, e^{px}} + \dfrac{1}{p}\left(-\dfrac{\cos x}{p \, e^{px}} - \dfrac{1}{p} \int \sin x \, e^{-px} \, dx \right)$

$f(x) =$
$\underline{\sin x}$

$\int \sin x \, e^{-px} \, dx$

$= -\dfrac{\sin x}{p \, e^{px}} - \dfrac{\cos x}{p^2 \, e^{px}} - \dfrac{1}{p^2} \int \sin x \, e^{-px} \, dx$

$\int \sin x \, e^{-px} \, dx + \dfrac{1}{p^2} \int \sin x \, e^{-px} \, dx$

$= -\dfrac{\sin x}{p \, e^{px}} - \dfrac{\cos x}{p^2 \, e^{px}}$

$\int \sin x \, e^{-px} \, dx \left(1 + \dfrac{1}{p^2} \right)$

$= -\dfrac{p \sin x}{p^2 \, e^{px}} - \dfrac{\cos x}{p^2 \, e^{px}}$

$\int \sin x \, e^{-px} \, dx \left(\dfrac{p^2}{p^2} + \dfrac{1}{p^2} \right)$

$= -\left(\dfrac{p \sin x}{p^2 \, e^{px}} + \dfrac{\cos x}{p^2 \, e^{px}} \right)$

$\int \sin x \, e^{-px} \, dx \left(\dfrac{p^2 + 1}{p^2} \right)$

$= -\left(\dfrac{p \sin x + \cos x}{p^2 \, e^{px}} \right)$

(cont)

$$\int \sin x \; e^{-px} \; dx$$

$$= \left(\frac{p^2}{p^2+1} \right) \left[-\left(\frac{p \sin x + \cos x}{p^2 \, e^{px}} \right) \right]$$

$f(x) =$
$\underline{\sin x}$

$$\int \sin x \; e^{-px} \; dx$$

$$= - \; \frac{p \sin x + \cos x}{(p^2+1) \, e^{px}}$$

Chapt 7 The Laplace Transformation of
Trigonometric Functions

(cont)

$$\int_0^{+\infty} \sin x \; e^{-px} \, dx$$

$$= - \left. \frac{p \sin x + \cos x}{(p^2+1) \; e^{px}} \right|_0^{+\infty}$$

$b(x) =$

$\sin x \qquad$
$$= - 0 + \frac{p \sin 0 + \cos 0}{(p^2+1) \; e^0}$$

$$= 0 + \left[\frac{0 + 1}{(p^2+1) \; 1} \right]$$

$$= \frac{1}{p^2+1}$$

note

$\sin x$ is bounded

$|\sin x| \le 1 \qquad$ on $\quad -1 \le \sin x \le +1$

and

$\cos x$ is bounded

$|\cos x| \le 1 \qquad$ on $\quad -1 \le \cos x \le +1$

II A

The Laplace Transformation of

Cent 7 Trigonometric Functions

② use the definition of the Laplace transformation to find the Laplace transform of

$f(x) = \cos x$

$f(x) = \cos x$

$$L[f(x)] = \int_0^{+\infty} f(x)\, e^{-\rho x}\, dx$$

$$L[\cos x] = \int_0^{+\infty} \cos x\; e^{-\rho x}\, dx$$

II B

Chpt 7 — The Laplace Transforms Of Trigonometric Functions — Prob

(2 cont)

$$\int \cos x \; e^{-px} \, dx$$

$f(x) = \cos x$

$$\int u \, dv = uv - \int v \, du$$

let $u = \cos x$ $dv = e^{-px} \, dx$

$du = -\sin x \, dx$ $\int dv = \int e^{-px} \, dx$

$$v = -\frac{1}{p} e^{-px}$$

$$\int \cos x \; e^{-px} \, dx$$

$$= \cos x \left(-\frac{1}{p} e^{-px} \right) - \int \left(-\frac{1}{p} e^{-px} \right) \left(-\sin x \, dx \right)$$

$$= -\frac{\cos x}{p \, e^{px}} - \frac{1}{p} \int \sin x \; e^{-px} \, dx$$

(71)

The Laplace Transform of

Ch pt 7   Trigonometric Functions

(2 cont) $\int \sin x \, e^{-px} \, dx$

$\int u \, dv = uv - \int v \, du$

let $u = \sin x$ $\qquad$ $dv = e^{-px} \, dx$

$g(x) =$
$\cos x$

$du = \cos x \, dx$ $\qquad$ $\int dv = \int e^{-px} \, dx$

$v = -\frac{1}{p} e^{-px}$

$\int \sin x \, e^{-px} \, dx$

$= \sin x \left(-\frac{1}{p} e^{-px} \right) - \int \left(-\frac{1}{p} e^{-px} \right) \cos x \, dx$

$= -\frac{\sin x}{p \, e^{px}} + \frac{1}{p} \int \cos x \, e^{-px} \, dx$

chrt 7   The Laplace Transformation of
Trigonometric Functions

(2 cont)

$$\int \cos x \; e^{-px} \, dx$$

$$= -\frac{\cos x}{p \, e^{px}} - \frac{1}{p}\left(-\frac{\sin x}{p \, e^{px}} + \frac{1}{p}\int \cos x \; e^{-px} \, dx\right)$$

$$6(x) = \\ \cos x \Bigg\} = \int \cos x \; e^{-px} \, dx$$

$$= -\frac{\cos x}{p \, e^{px}} + \frac{\sin x}{p^2 \, e^{px}} - \frac{1}{p^2}\int \cos x \; e^{-px} \, dx$$

$$\int \cos x \; e^{-px} \, dx + \frac{1}{p^2}\int \cos x \; e^{-px} \, dx$$

$$= \frac{\sin x}{p^2 \, e^{px}} - \frac{\cos x}{p \, e^{px}}$$

$$\int \cos x \; e^{-px} \, dx \left(1 + \frac{1}{p^2}\right)$$

$$= \frac{\sin x}{p^2 \, e^{px}} - \frac{p \cos x}{p^2 \, e^{px}}$$

$$\int \cos x \; e^{-px} \, dx \left(\frac{p^2}{p^2} + \frac{1}{p^2}\right)$$

$$= \frac{\sin x - p \cos x}{p^2 \, e^{px}}$$

$$\int \cos x \; e^{-px} \, dx \left(\frac{p^2 + 1}{p^2}\right)$$

$$= \frac{\sin x - p \cos x}{p^2 \, e^{px}}$$

(73)

The Laplace Transform of Trigonometric Functions

Art 7

$$\int \cos x \; e^{-px} \, dx$$

$$= \left(\frac{p^2}{p^2+1} \right) \left(\frac{\sin x - p\cos x}{p^2 \, e^{px}} \right)$$

$$\frac{b(x) =}{\cos x} \quad \int \cos x \; e^{-px} \, dx$$

$$= \frac{\sin x - p\cos x}{(p^2+1) \, e^{px}}$$

II F

Chpt 7 The Laplace Transform Of Trigonometric Functions

(cont)

$$\int_0^{+\infty} \cos x \; e^{-px} \, dx$$

$$= \left. \frac{\sin x - p\cos x}{(p^2+1)\, e^{px}} \right|_0^{+\infty}$$

$G(x) = \cos x$

$$= 0 - \left[\frac{\sin 0 - p\cos 0}{(p^2+1)\, e^0} \right]$$

$$= - \left[\frac{0 - p(1)}{(p^2+1)(1)} \right]$$

$$= - \left(- \frac{p}{p^2+1} \right)$$

$$= \frac{p}{p^2+1}$$

note

$\sin x$ is bounded

$|\sin x| \leq 1$ on $-1 \leq \sin x \leq +1$

and

$\cos x$ is bounded

$|\cos x| \leq 1$ on $-1 \leq \cos x \leq +1$

The Laplace Transformation of

Chapt 7. Trigonometric Functions

① use the definition of the Laplace transformation to find the Laplace transform of

$$f(x) = \sin kx$$

$$f(x) = \sin kx$$

$$L[f(x)] = \int_0^{+\infty} f(x)\, e^{-px}\, dx$$

$$L[\sin kx] = \int_0^{+\infty} \sin kx \; e^{-px}\, dx$$

The Laplace Transform of
Chpt 7 Trigonometric Functions

(I cont)

$$\int \sin kx \ e^{-px} \ dx$$

$$\int u \ dv = uv - \int v \ du$$

$$\text{let } u = \sin kx$$

$b(x) =$
$\sin kx$

$$\frac{du}{dx} = \cos kx \ (k)$$

$$dv = e^{-px} \ dx$$

$$\int dv = \int e^{-px} \ dx$$

$$v = -\frac{1}{p} e^{-px}$$

$$\frac{du}{dx} = k \cos kx$$

$$du = k \cos kx \ dx$$

$$\int \sin kx \ e^{-px} \ dx$$

$$= \sin kx \left(-\frac{1}{p} e^{-px}\right) - \int \left(-\frac{1}{p} e^{-px}\right) k \cos kx \ dx$$

$$= -\frac{\sin kx}{p \, e^{px}} + \frac{k}{p} \int \cos kx \ e^{-px} \ dx$$

IC (77) Prob

The Laplace Transformation Of
Chpt 7 Trigonometric Functions

$$\int \cos kx \; e^{-px} \, dx$$

$$\int u \, dv = uv - \int v \, du$$

let $u = \cos kx$

$$\frac{du}{dx} = -\sin kx \, (k)$$

$$\frac{du}{dx} = -k \sin kx$$

$$du = -k \sin kx \, dx$$

$f(x) = \sin kx$

$dv = e^{-px} \, dx$

$$\int dv = \int e^{-px} \, dx$$

$$v = -\frac{1}{p} e^{-px}$$

$$\int \cos kx \; e^{-px} \, dx$$

$$= \cos kx \left(-\frac{1}{p} e^{-px}\right) - \int \left(-\frac{1}{p} e^{-px}\right) \left(-k \sin kx \, dx\right)$$

$$= -\frac{\cos kx}{p \, e^{px}} - \frac{k}{p} \int \sin kx \; e^{-px} \, dx$$

The Laplace Transformation of Trigonometric Functions

Ch 7

(cont) $\int \sin kx \; e^{-px} \, dx$

$$= -\frac{\sin kx}{p \, e^{px}} + \frac{k}{p}\left(-\frac{\cos kx}{p \, e^{px}} - \frac{k}{p}\int \sin kx \; e^{-px} \, dx\right)$$

$f(x) =$
$\sin kx$ $\int \sin kx \; e^{-px} \, dx$

$$= -\frac{\sin kx}{p \, e^{px}} - \frac{k\cos kx}{p^2 \, e^{px}} - \frac{k^2}{p^2}\int \sin kx \; e^{-px} \, dx$$

$$\int \sin kx \; e^{-px} \, dx + \frac{k^2}{p^2}\int \sin kx \; e^{-px} \, dx$$

$$= -\frac{\sin kx}{p \, e^{px}} - \frac{k\cos kx}{p^2 \, e^{px}}$$

$$\int \sin kx \; e^{-px} \, dx \left(1 + \frac{k^2}{p^2}\right)$$

$$= -\left(\frac{\sin kx}{p \, e^{px}} + \frac{k\cos kx}{p^2 \, e^{px}}\right)$$

$$\int \sin kx \; e^{-px} \, dx \left(\frac{p^2}{p^2} + \frac{k^2}{p^2}\right)$$

$$= -\left(\frac{p\sin kx}{p^2 \, e^{px}} + \frac{k\cos kx}{p^2 \, e^{px}}\right)$$

$$\int \sin kx \; e^{-px} \, dx \left(\frac{p^2 + k^2}{p^2}\right)$$

$$= -\left(\frac{p\sin kx + k\cos kx}{p^2 \, e^{px}}\right)$$

The Laplace Transformation Of
Trigonometric Functions

Chapt 7

$$\int \sin kx \; e^{-px} \, dx$$

$$= -\left(\frac{p^2}{p^2 + k^2} \right) \left(\frac{p \sin kx + k \cos kx}{p^2 \, e^{px}} \right)$$

$G(x) =$
$\underline{\sin kx} \quad \int \sin kx \; e^{-px} \, dx$

$$= - \; \frac{p \sin kx + k \cos kx}{(p^2 + k^2) \, e^{px}}$$

The Laplace Transform of Trigonometric Functions

Chapt 7

(cont) $\int_0^{+\infty} \sin kx \; e^{-px} \; dx$

$$= - \frac{p \sin kx + k \cos kx}{(p^2 + k^2) \; e^{px}} \Big|_0^{+\infty}$$

$$= - 0 + \frac{p \sin 0 + k \cos 0}{(p^2 + k^2) \; e^0}$$

$\int kx = \frac{p(0) + k(1)}{(p^2 + k^2)(1)}$

$$= \frac{k}{p^2 + k^2}$$

note

$\sin kx$ is bounded

$|\sin kx| \leq 1$ on $-1 \leq \sin kx \leq +1$

and

$\cos kx$ is bounded

$|\cos kx| \leq 1$ on $-1 \leq \cos kx \leq +1$

The Laplace Transformation of

② Chapt 7 Trigonometric Functions

use the definition of the Laplace transformation
to find the Laplace transform of

$b(x) = \cos kx$

$b(x) = \cos kx$

$$L[b(x)] = \int_0^{+\infty} b(x)\, e^{-px}\, dx$$

$$L[\cos kx] = \int_0^{+\infty} \cos kx\; e^{-px}\, dx$$

The Laplace Transform of
Chapt 7   Trigonometric Functions

(2 cont)

$$\int \cos kx \, e^{-px} \, dx$$

$$\int u \, dv = uv - \int v \, du$$

let $u = \cos kx$ $dv = e^{-px} \, dx$

$f(x) =$ $\dfrac{du}{dx} = -\sin kx \, (k)$ $\int dv = \int e^{-px} \, dx$

$\cos kx$ $v = -\dfrac{1}{p} e^{-px}$

$$\dfrac{du}{dx} = -k \sin kx$$

$$du = -k \sin kx \, dx$$

$$\int \cos kx \, e^{-px} \, dx$$

$$= \cos kx \left(-\frac{1}{p} e^{-px} \right) - \int \left(-\frac{1}{p} e^{-px} \right) \left(-k \sin kx \, dx \right)$$

$$= - \frac{\cos kx}{p \, e^{px}} - \frac{k}{p} \int \sin kx \, e^{-px} \, dx$$

The Laplace Transformation Of Trigonometric Functions

(2 Cont)

$$\int \sin KX \ e^{-px} \ dx$$

$$\int u \ dv = uv - \int v \ du$$

let $u = \sin KX$

$\dfrac{du}{dx} = \cos KX \ (K)$

$f(x) = \cos KX$

$\dfrac{du}{dx} = K \cos KX$

$du = K \cos KX \ dx$

$dv = e^{-px} \ dx$

$\int dv = \int e^{-px} \ dx$

$v = -\dfrac{1}{p} \ e^{-px}$

$$\int \sin KX \ e^{-px} \ dx$$

$$= \sin KX \left[-\frac{1}{p} e^{-px} \right] - \int \left(-\frac{1}{p} e^{-px} \right) K \cos KX \ dx$$

$$= - \frac{\sin KX}{p \ e^{px}} + \frac{K}{p} \int \cos KX \ e^{-px} \ dx$$

$$\underline{\text{Chapt 7} \quad \overline{\text{The Laplace Transformation of}}}$$
$$\overline{\text{Trigonometric Functions}}$$

(2 cont) $\int \cos Kx \; e^{-px} \, dx$

$$= - \frac{\cos Kx}{p \, e^{px}} - \frac{K}{p} \left(- \frac{\sin Kx}{p \, e^{px}} + \frac{K}{p} \int \cos Kx \; e^{-px} \, dx \right)$$

$\underline{f(x) =}$
$\underline{\cos Kx}$ $\int \cos Kx \; e^{-px} \, dx$

$$= - \frac{\cos Kx}{p \, e^{px}} + \frac{K \sin Kx}{p^2 \, e^{px}} - \frac{K^2}{p^2} \int \cos Kx \; e^{-px} \, dx$$

$$\int \cos Kx \; e^{-px} \, dx + \frac{K^2}{p^2} \int \cos Kx \; e^{-px} \, dx$$

$$= \frac{K \sin Kx}{p^2 \, e^{px}} - \frac{\cos Kx}{p \, e^{px}}$$

$$\int \cos Kx \; e^{-px} \, dx \left(1 + \frac{K^2}{p^2} \right)$$

$$= \frac{K \sin Kx}{p^2 \, e^{px}} - \frac{p \cos Kx}{p^2 \, e^{px}}$$

$$\int \cos Kx \; e^{-px} \, dx \left(\frac{p^2}{p^2} + \frac{K^2}{p^2} \right)$$

$$= \frac{K \sin Kx - p \cos Kx}{p^2 \, e^{px}}$$

$$\int \cos Kx \; e^{-px} \, dx \left(\frac{p^2 + K^2}{p^2} \right)$$

$$= \frac{K \sin Kx - p \cos Kx}{p^2 \, e^{px}}$$

II E

(85)

(2cont)

$$\int \cos kx \, e^{-px} \, dx$$

$$= \left(\frac{p^2}{p^2 + k^2} \right) \left(\frac{k \sin kx - p \cos kx}{p^2 \, e^{px}} \right)$$

$f(x) = \cos kx$

$$\int \cos kx \, e^{-px} \, dx$$

$$= \frac{k \sin kx - p \cos kx}{(p^2 + k^2) \, e^{px}}$$

The Laplace Transformation of Chapter 7 Trigonometric Functions

$$\int_0^{+\infty} \cos kx \, e^{-px} \, dx$$

$$= \left. \frac{k \sin kx - p \cos kx}{(p^2 + k^2) \, e^{px}} \right|_0^{+\infty}$$

$$f(x) = \atop \cos kx \qquad = 0 - \left[\frac{k \sin 0 - p \cos 0}{(p^2 + k^2) \, e^0} \right]$$

$$= - \left[\frac{k(0) - p(1)}{(p^2 + k^2)(1)} \right]$$

$$= - \left(\frac{0 - p}{p^2 + k^2} \right)$$

$$= \frac{p}{p^2 + k^2}$$

note

$\sin kx$ is bounded

$|\sin kx| \leq 1$ on $-1 \leq \sin kx \leq +1$

and

$\cos kx$ is bounded

$|\cos kx| \leq 1$ on $-1 \leq \cos kx \leq +1$

The Laplace Transformation Of
Chapt 8 Hyperbolic Functions
The Laplace Transformation Of
Hyperbolic Functions

① In this chapter we shall use
the definition of the Laplace transformation

$$L[f(x)] = \int_0^{+\infty} f(x) \, e^{-px} \, dx$$

to derive the Laplace transform of
each of the following hyperbolic functions

$$f(x) = \sinh x$$
$$f(x) = \cosh x$$

$$f(x) = \sinh kx \qquad \text{where } k \text{ is a constant}$$
$$f(x) = \cosh kx \qquad \text{where } k \text{ is a constant}$$

The Laplace Transformation Of

Unit 8 Hyperbolic Functions

2) the results we want to obtain are

$f(x) = \sinh x$

$L[\sinh x] = \dfrac{1}{p^2 - 1}$

$f(x) = \cosh x$

$L[\cosh x] = \dfrac{p}{p^2 - 1}$

$f(x) = \sinh kx$

$L[\sinh kx] = \dfrac{k}{p^2 - k^2}$

$f(x) = \cosh kx$

$L[\cosh kx] = \dfrac{p}{p^2 - k^2}$

I A

Chapt 8   The Laplace Transformation of Hyperbolic Functions

① Find the Laplace transform of $f(x) = \sinh x$

$$L[f(x)] = \int_0^{+\infty} f(x) \, e^{-px} \, dx$$

$f(x) = \sinh x$

$$L[\sinh x] = \int_0^{+\infty} \sinh x \; e^{-px} \, dx$$

$$\sinh x = \frac{e^x - e^{-x}}{2}$$

$$L[\sinh x] = \int_0^{+\infty} \tfrac{1}{2}(e^x - e^{-x}) \, e^{-px} \, dx$$

$$= \int_0^{+\infty} \left(\tfrac{1}{2} e^x \cdot e^{-px} - \tfrac{1}{2} e^{-x} \cdot e^{-px} \right) dx$$

$$= \int_0^{+\infty} \left(\tfrac{1}{2} e^{x-px} - \tfrac{1}{2} e^{-x-px} \right) dx$$

$$= \int_0^{+\infty} \tfrac{1}{2} e^{(1-p)x} - \tfrac{1}{2} e^{(-1-p)x} \, dx$$

$$= \int_0^{+\infty} \tfrac{1}{2} e^{-(p-1)x} - \tfrac{1}{2} e^{-(1+p)x} \, dx$$

$$= \int_0^{+\infty} \tfrac{1}{2} e^{-(p-1)x} - \tfrac{1}{2} e^{-(p+1)x} \, dx$$

$$= \left\{ \tfrac{1}{2} \left[-\left(\frac{1}{p-1} \right) \right] e^{-(p-1)x} - \tfrac{1}{2} \left[-\left(\frac{1}{p+1} \right) \right] e^{-(p+1)x} \right\} \Big|_0^{+\infty}$$

$$= \left[- \frac{1}{2(p-1) \, e^{(p-1)x}} + \frac{1}{2(p+1) \, e^{(p+1)x}} \right] \Big|_0^{+\infty}$$

The Laplace Transformation of
Chapt 8   Hyperbolic Functions

$$
\text{cont} = \left[\frac{1}{2(p+1)\,e^{(p+1)x}} - \frac{1}{2(p-1)\,e^{(p-1)x}} \right] \Big|_{0}^{+\infty}
$$

$$
= (0 - 0) - \left[\frac{1}{2(p+1)\,e^{0}} - \frac{1}{2(p-1)\,e^{0}} \right]
$$

$$
= - \left[\frac{1}{2(p+1)(1)} - \frac{1}{2(p-1)(1)} \right]
$$

$$
= - \left[\frac{1}{2(p+1)} - \frac{1}{2(p-1)} \right]
$$

$$
= - \frac{1}{2(p+1)} + \frac{1}{2(p-1)}
$$

$$
= \frac{1}{2(p-1)} - \frac{1}{2(p+1)}
$$

$$
= \frac{1}{2(p-1)} \frac{p+1}{p+1} - \frac{1}{2(p+1)} \frac{p-1}{p-1}
$$

$$
= \frac{p+1 - p+1}{2(p-1)(p+1)}
$$

$$
= \frac{2}{2(p-1)(p+1)}
$$

$$
= \frac{1}{(p-1)(p+1)}
$$

$$
= \frac{1}{p^2 - 1}
$$

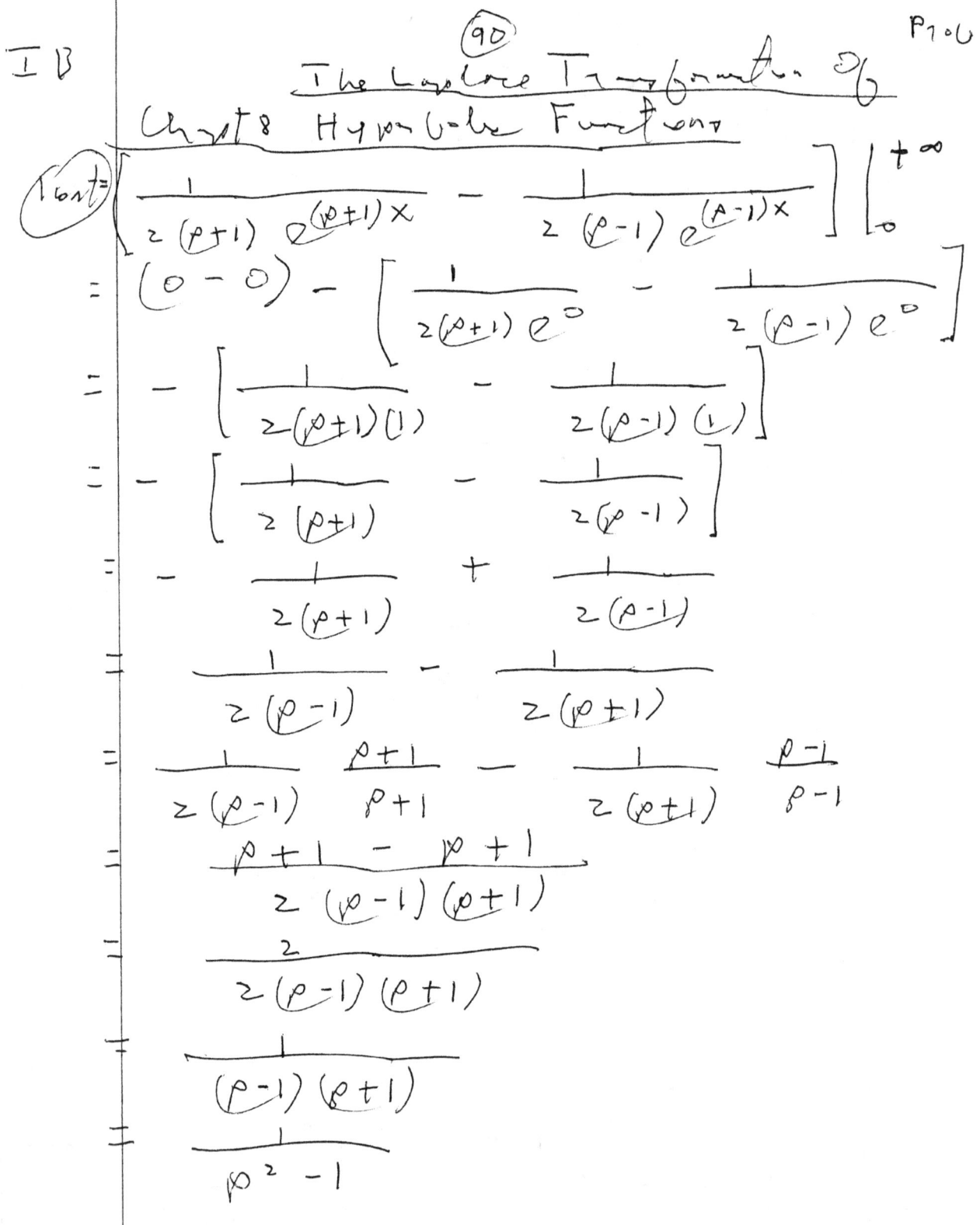

(91)

The Laplace Transformation of

Chapter 8   Hyperbolic Functions

② find the Laplace transform for $f(x) = \cosh x$

$$L[f(x)] = \int_0^{+\infty} f(x)\, e^{-px}\, dx$$

$f(x) = \cosh x$

$$L[\cosh x] = \int_0^{+\infty} \cosh x\; e^{-px}\, dx$$

$$\cosh x = \frac{e^x + e^{-x}}{2}$$

$$L[\cosh x] = \int_0^{+\infty} \frac{1}{2}(e^x + e^{-x})\, e^{-px}\, dx$$

$$= \int_0^{+\infty} \left(\frac{1}{2} e^x \cdot e^{-px} + \frac{1}{2} e^{-x} \cdot e^{-px}\right) dx$$

$$= \int_0^{+\infty} \left(\frac{1}{2} e^{x-px} + \frac{1}{2} e^{-x-px}\right) dx$$

$$= \int_0^{+\infty} \left[\frac{1}{2} e^{(1-p)x} + \frac{1}{2} e^{(-1-p)x}\right] dx$$

$$= \int_0^{+\infty} \left[\frac{1}{2} e^{-(p-1)x} + \frac{1}{2} e^{-(1+p)x}\right] dx$$

$$= \int_0^{+\infty} \left[\frac{1}{2} e^{-(p-1)x} + \frac{1}{2} e^{-(p+1)x}\right] dx$$

$$= \left\{\frac{1}{2}\left[-\left(\frac{1}{p-1}\right)\right] e^{-(p-1)x} + \frac{1}{2}\left[-\left(\frac{1}{p+1}\right)\right] e^{-(p+1)x}\right\}\Bigg|_0^{+\infty}$$

$$= \left[-\frac{1}{2(p-1)\, e^{(p-1)x}} - \frac{1}{2(p+1)\, e^{(p+1)x}}\right]\Bigg|_0^{+\infty}$$

$$\boxed{92}$$

The Laplace Transformation of
Chapt 8   The Hyperbolic Functions

$$= \left(-\frac{1}{\infty} - \frac{1}{\infty} \right) - \left[- \frac{1}{2(p-1)\,e^{0}} - \frac{1}{2(p+1)\,e^{0}} \right]$$

$$= \left(-0 - 0 \right) + \frac{1}{2(p-1)(1)} + \frac{1}{2(p+1)(1)}$$

$$= 0 + \frac{1}{2(p-1)} + \frac{1}{2(p+1)}$$

$$= \frac{1}{2(p-1)} + \frac{1}{2(p+1)}$$

$$= \frac{1}{2(p-1)} \frac{p+1}{p+1} + \frac{1}{2(p+1)} \frac{p-1}{p-1}$$

$$= \frac{p+1 + p-1}{2(p-1)(p+1)}$$

$$= \frac{2p}{2(p-1)(p+1)}$$

$$= \frac{p}{(p-1)(p+1)}$$

$$= \frac{p}{p^{2}-1}$$

The Laplace Transform of

Chapt 8 Hyperbolic Functions

① find the Laplace transform of $f(x) = \sinh Kx$

$$L[f(x)] = \int_0^{+\infty} f(x) \, e^{-px} \, dx$$

$f(x) = \sinh KX$

$$L[\sinh Kx] = \int_0^{+\infty} \sinh Kx \, e^{-px} \, dx$$

$$\sinh x = \frac{e^x - e^{-x}}{2}$$

$$\sinh Kx = \frac{e^{Kx} - e^{-Kx}}{2}$$

$$L[\sinh Kx] = \int_0^{+\infty} \frac{1}{2}(e^{Kx} - e^{-Kx}) \, e^{-px} \, dx$$

$$= \int_0^{+\infty} (\frac{1}{2} e^{Kx} \cdot e^{-px} - \frac{1}{2} e^{-Kx} \cdot e^{-px}) \, dx$$

$$= \int_0^{+\infty} (\frac{1}{2} e^{Kx - px} - \frac{1}{2} e^{-Kx - px}) \, dx$$

$$= \int_0^{+\infty} [\frac{1}{2} e^{(K-p)x} - \frac{1}{2} e^{(-K-p)x}] \, dx$$

$$= \int_0^{+\infty} [\frac{1}{2} e^{-(p-K)x} - \frac{1}{2} e^{-(K+p)x}] \, dx$$

$$= \int_0^{+\infty} [\frac{1}{2} e^{-(p-K)x} - \frac{1}{2} e^{-(p+K)x}] \, dx$$

$$= \left\{ [\frac{1}{2}[-(\frac{1}{p-K})] e^{-(p-K)x} - \frac{1}{2}[-(\frac{1}{p+K})] e^{-(p+K)x} \right\}\Big|_0^+$$

The Laplace Transformation of
chapt 8   Hyperbolic Functions

$$\boxed{\text{Cont}}\left[-\frac{1}{2(p-k)}e^{(p-k)x} + \frac{1}{2(p+k)}e^{(p+k)x}\right]\Bigg|_0^{+\infty}$$

$$= \left[\frac{1}{2(p+k)}e^{(p+k)x} - \frac{1}{2(p-k)}e^{(p-k)x}\right]\Bigg|_0^{+\infty}$$

$$= \left(\frac{1}{\infty} - \frac{1}{\infty}\right) - \left[\frac{1}{2(p+k)e^0} - \frac{1}{2(p-k)e^0}\right]$$

$$= (0-0) - \left[\frac{1}{2(p+k)(1)} - \frac{1}{2(p-k)(1)}\right]$$

$$= -\left[\frac{1}{2(p+k)} - \frac{1}{2(p-k)}\right]$$

$$= \frac{1}{2(p-k)} - \frac{1}{2(p+k)}$$

$$= \frac{1}{2(p-k)}\frac{p+k}{p+k} - \frac{1}{2(p+k)}\frac{p-k}{p-k}$$

$$= \frac{p+k - p + k}{2(p-k)(p+k)}$$

$$= \frac{2k}{2(p-k)(p+k)}$$

$$= \frac{k}{(p-k)(p+k)}$$

$$= \frac{k}{p^2 - k^2}$$

The Laplace Transformation of

Unit 8 Hyperbolic Functions

2) Find the Laplace transform of $f(x) = \cosh KX$

$$L(f(x)) = \int_0^{+\infty} f(x) \, e^{-px} \, dx$$

$f(x) = \cosh KX$

$$L[\cosh KX] = \int_0^{+\infty} \cosh KX \; e^{-px} \, dx$$

$$\cosh x = \frac{e^x + e^{-x}}{2}$$

$$\cosh KX = \frac{e^{KX} + e^{-KX}}{2}$$

$$L[\cosh KX] = \int_0^{+\infty} \tfrac{1}{2}\left(e^{KX} + e^{-KX}\right) e^{-px} \, dx$$

$$= \int_0^{+\infty} \left(\tfrac{1}{2} e^{KX} \cdot e^{-px} + \tfrac{1}{2} e^{-KX} \cdot e^{-px}\right) dx$$

$$= \int_0^{+\infty} \left(\tfrac{1}{2} e^{KX-px} + \tfrac{1}{2} e^{-KX-px}\right) dx$$

$$= \int_0^{+\infty} \left[\tfrac{1}{2} e^{(K-p)x} + \tfrac{1}{2} e^{(-K-p)x}\right] dx$$

$$= \int_0^{+\infty} \left[\tfrac{1}{2} e^{-(p-K)x} + \tfrac{1}{2} e^{-(K+p)x}\right] dx$$

$$= \int_0^{+\infty} \left[\tfrac{1}{2} e^{-(p-K)x} + \tfrac{1}{2} e^{-(p+K)x}\right] dx$$

$$= \left\{\tfrac{1}{2}\left[-\left(\frac{1}{p-K}\right)\right] e^{-(p-K)x} + \tfrac{1}{2}\left[-\left(\frac{1}{p+K}\right)\right] e^{-(p+K)x}\right\}\Bigg|_0^{+}$$

The Laplace Transformation Of
Unto Hyperbolic Functions

(2 cont) $= \left(- \dfrac{1}{2(p-k) \, e^{(p-k)x}} - \dfrac{1}{2(p+k) \, e^{(p+k)x}} \right) \Big|_0^{+\infty}$

$= \left(- \dfrac{1}{\infty} - \dfrac{1}{\infty} \right) - \left[- \dfrac{1}{2(p-k) \, e^0} - \dfrac{1}{2(p+k) \, e^0} \right]$

$= (-0-0) + \dfrac{1}{2(p-k)(1)} + \dfrac{1}{2(p+k)(1)}$

$= 0 + \dfrac{1}{2(p-k)} + \dfrac{1}{2(p+k)}$

$= \dfrac{1}{2(p-k)} + \dfrac{1}{2(p+k)}$

$= \dfrac{1}{2(p-k)} \cdot \dfrac{p+k}{p+k} + \dfrac{1}{2(p+k)} \cdot \dfrac{p-k}{p-k}$

$= \dfrac{p+k + p - k}{2(p-k)(p+k)}$

$= \dfrac{2p}{2(p-k)(p+k)}$

$= \dfrac{p}{(p-k)(p+k)}$

$= \dfrac{p}{p^2 - k^2}$

Chapt 9 Making A Table Of Laplace Transform

Making A Table of Laplace Transforms

(1) we have used the definition of
the Laplace transformation

$$L[b(x)] = \int_0^{+\infty} b(x) \, e^{-px} \, dx$$

to find the Laplace transforms of
the following functions

Chapt 9 Making A Table Of Laplace Transform

$f(x) = 0$

$f(x) = 1$

$f(x) = K$ where K is a constant

$f(x) = x$

$f(x) = x^2$

$f(x) = x^3$

$f(x) = x^4$

$\vdots$

$f(x) = x^n$

$f(x) = e^x$

$f(x) = e^{Kx}$ where K is a constant

[$f(x) = \sin x$

   $f(x) = \cos x$

[$f(x) = \sin Kx$ where K is a constant

   $f(x) = \cos Kx$ where K is a constant

[$f(x) = \sinh x$

   $f(x) = \cosh x$

[$f(x) = \sinh Kx$ where K is a constant

   $f(x) = \cosh Kx$ where K is a constant

(2) we obtained the following results

Chapter Making A Table of Laplace Transform

function $b(x)$ $\xrightarrow{L}$	Laplace transform $F(p)$
$b(x) = 0$	$F(p) = 0$
$b(x) = 1$	$F(p) = \dfrac{1}{p}$
$b(x) = K$	$F(p) = \dfrac{K}{p}$
$b(x) = x$	$F(p) = \dfrac{1!}{p^{1+1}}$
$b(x) = x^2$	$F(p) = \dfrac{2!}{p^{2+1}}$
$b(x) = x^3$	$F(p) = \dfrac{3!}{p^{3+1}}$
$b(x) = x^4$	$F(p) = \dfrac{4!}{p^{4+1}}$
$b(x) = x^n$	$F(p) = \dfrac{n!}{p^{n+1}}$
$b(x) = e^x$	$F(p) = \dfrac{1}{p-1}$
$b(x) = e^{Kx}$	$F(p) = \dfrac{1}{p-K}$

$\downarrow$ intro

Chapt 9 Mabry A Table Of Laplace Transform

$$\text{function } f(x) \xrightarrow{\quad L \quad} \text{Laplace transform } F(p)$$

$f(x) = \sin x$ $\qquad F(p) = \dfrac{1}{p^2 + 1}$

$f(x) = \cos x$ $\qquad F(p) = \dfrac{p}{p^2 + 1}$

$f(x) = \sin kx$ $\qquad F(p) = \dfrac{k}{p^2 + k^2}$

$f(x) = \cos kx$ $\qquad F(p) = \dfrac{p}{p^2 + k^2}$

$f(x) = \sinh x$ $\qquad F(p) = \dfrac{1}{p^2 - 1}$

$f(x) = \cosh x$ $\qquad F(p) = \dfrac{p}{p^2 - 1}$

$f(x) = \sinh kx$ $\qquad F(p) = \dfrac{k}{p^2 - k^2}$

$f(x) = \cosh kx$ $\qquad F(p) = \dfrac{p}{p^2 - k^2}$

(101)

Ch. 9 Making A Table Of Laplace Transforms

(3) we may abbreviate these results
to make a more compact table
that still includes all of this information

(102)

Chapter Making A Table of Laplace Transform

function $G(X)$ $\xrightarrow{L}$	Laplace transform $F(p)$
0	0
1	$\dfrac{1}{p}$
K	$\dfrac{K}{p}$
X^n $n = 0, 1, 2, 3, \ldots$	$\dfrac{n!}{p^{n+1}}$
e^{KX}	$\dfrac{1}{p - K}$
$\sin KX$	$\dfrac{K}{p^2 + K^2}$
$\cos KX$	$\dfrac{p}{p^2 + K^2}$
$\sinh KX$	$\dfrac{K}{p^2 - K^2}$
$\cosh KX$	$\dfrac{p}{p^2 - K^2}$

note
K is a constant

chapt 10 Using The Table To Find Laplace Transforms

Using The Table To Find Laplace Transforms
Of Elementary Functions

① we may use the following table as mode
to find the Laplace transform
of specify elementary functions.

② the elementary functions may be categoriyed
into the following groups

① constants
② polynomials
③ exponential
④ trigonometric
⑤ hyperbolic

Chapt 10 Using The Table To Find Laplace Transform

The function $f(x)$ $\xrightarrow{L}$	the Laplace Transform $F(p)$
0	0
1	$\dfrac{1}{p}$
K	$\dfrac{K}{p}$
x^n $\quad n = 0, 1, 2, 3, \ldots$	$\dfrac{n!}{p^{n+1}}$
e^{Kx}	$\dfrac{1}{p-K}$
$\sin Kx$	$\dfrac{K}{p^2 + K^2}$
$\cos Kx$	$\dfrac{p}{p^2 + K^2}$
$\sinh Kx$	$\dfrac{K}{p^2 - K^2}$
$\cosh Kx$	$\dfrac{p}{p^2 - K^2}$

chapt 10 Using The Table To Find Laplace Transforms

ex   find the Laplace transform of the function $f(x) = 0$

$$f(x) = 0$$

$$L[K] = \frac{K}{p}$$

$$L[0] = \frac{0}{p}$$

$$L[0] = 0$$

ex   find the Laplace transform of the function $f(x) = 1$

$$f(x) = 1$$

$$L[K] = \frac{K}{p}$$

$$L[1] = \frac{1}{p}$$

Chapter 10 Using The Table To Find Laplace Transform

Ex. Find the Laplace transform of the function $b(x) = 2$

$b(x) = 2$

$L[K] = \dfrac{K}{p}$

$L[2] = \dfrac{2}{p}$

Ex. Find the Laplace transform of the function $b(x) = 3$

$b(x) = 3$

$L[K] = \dfrac{K}{p}$

$L[3] = \dfrac{3}{p}$

(107)

Ch. 10 Using The Table To Find Laplace Transform

ex | find the Laplace transform of $f(x) = -1$

$f(x) = -1$

$L[K] = \dfrac{K}{p}$

$L[-1] = \dfrac{-1}{p} \quad = \quad -\dfrac{1}{p}$

ex | find the Laplace transform of $f(x) = -2$

$f(x) = -2$

$L[K] = \dfrac{K}{p}$

$L[-2] = \dfrac{-2}{p} \quad = \quad -\dfrac{2}{p}$

ex | find the Laplace transform of $f(x) = -3$

$f(x) = -3$

$L[K] = \dfrac{K}{p}$

$L[-3] = \dfrac{-3}{p} \quad = \quad -\dfrac{3}{p}$

$\mathcal{A}$ intro

(108)

Chapter 10 Using The Table To Find Laplace Transform

ex find the Laplace transform of $f(x) = \frac{1}{2}$

$f(x) = \frac{1}{2}$

$L[k] = \frac{k}{p}$

$L\left[\frac{1}{2}\right] = \frac{\frac{1}{2}}{p} = \frac{1}{2} \cdot \frac{1}{p} = \frac{1}{2p}$

ex find the Laplace transform of $f(x) = \frac{1}{3}$

$f(x) = \frac{1}{3}$

$L[k] = \frac{k}{p}$

$L\left[\frac{1}{3}\right] = \frac{\frac{1}{3}}{p} = \frac{1}{3} \cdot \frac{1}{p} = \frac{1}{3p}$

ex find the Laplace transform of $f(x) = \frac{5}{3}$

$f(x) = \frac{5}{3}$

$L[k] = \frac{k}{p}$

$L\left[\frac{5}{3}\right] = \frac{\frac{5}{3}}{p} = \frac{5}{3} \cdot \frac{1}{p} = \frac{5}{3p}$

Chapt 10 Using The Table To Find Laplace Transform

ex   Find the Laplace transform of $f(x) = -\frac{1}{2}$

$$f(x) = -\frac{1}{2}$$

$$L[K] = \frac{K}{p} = K \cdot \frac{1}{p}$$

$$L\left[-\frac{1}{2}\right] = -\frac{1}{2} \cdot \frac{1}{p} = -\frac{1}{2p}$$

ex   Find the Laplace transform of $f(x) = -\frac{1}{3}$

$$f(x) = -\frac{1}{3}$$

$$L[K] = \frac{K}{p} = K \cdot \frac{1}{p}$$

$$L\left[-\frac{1}{3}\right] = -\frac{1}{3} \cdot \frac{1}{p} = -\frac{1}{3p}$$

ex   Find the Laplace transform of $f(x) = -\frac{5}{3}$

$$f(x) = -\frac{5}{3}$$

$$L[K] = \frac{K}{p} = K \cdot \frac{1}{p}$$

$$L\left[-\frac{5}{3}\right] = -\frac{5}{3} \cdot \frac{1}{p} = -\frac{5}{3p}$$

Chapt 10 Using The Table To Find Laplace Transform polynomials

ex | find the Laplace transform of $f(x) = x$

$f(x) = x$

$$L[x^n] = \frac{n!}{p^{n+1}}$$

$$L[x^1] = \frac{1!}{p^{1+1}}$$

$$= \frac{1}{p^2}$$

ex | find the Laplace transform of $f(x) = x^2$

$f(x) = x^2$

$$L[x^n] = \frac{n!}{p^{n+1}}$$

$$L[x^2] = \frac{2!}{p^{2+1}}$$

$$= \frac{2 \cdot 1}{p^3}$$

$$= \frac{2}{p^3}$$

Chapt 10 Using The Table To Find Laplace Transform

ex   find the Laplace transform of $f(x) = x^3$

$$f(x) = x^3$$

$$L[x^n] = \frac{n!}{p^{n+1}}$$

$$L[x^3] = \frac{3!}{p^{3+1}}$$

$$= \frac{3 \cdot 2 \cdot 1}{p^4}$$

$$= \frac{6}{p^4}$$

ex   find the Laplace transform of $f(x) = x^4$

$$f(x) = x^4$$

$$L[x^n] = \frac{n!}{p^{n+1}}$$

$$L[x^4] = \frac{4!}{p^{4+1}}$$

$$= \frac{4 \cdot 3 \cdot 2 \cdot 1}{p^5}$$

$$= \frac{24}{p^5}$$

Chapt 10 Using The Table To Find Laplace Transform

exponential function

ex find the Laplace transform of $f(x) = e^x$

$f(x) = e^x$

$L[e^{kx}] = \dfrac{1}{p-k}$

$L[e^x] = \dfrac{1}{p-1}$

ex find the Laplace transform of $f(x) = e^{2x}$

$f(x) = e^{2x}$

$L[e^{kx}] = \dfrac{1}{p-k}$

$L[e^{2x}] = \dfrac{1}{p-2}$

ex find the Laplace transform of $f(x) = e^{3x}$

$f(x) = e^{3x}$

$L[e^{kx}] = \dfrac{1}{p-k}$

$L[e^{3x}] = \dfrac{1}{p-3}$

chpt 10 Using The Table To Find Laplace Transform

ex find the Laplace transform of $f(x) = e^{-x}$

$$f(x) = e^{-x}$$

$$L[e^{kx}] = \frac{1}{p-k}$$

$$L[e^{-x}] = \frac{1}{p-(-1)}$$

$$= \frac{1}{p+1}$$

ex find the Laplace transform of $f(x) = e^{-2x}$

$$f(x) = e^{-2x}$$

$$L[e^{kx}] = \frac{1}{p-k}$$

$$L[e^{-2x}] = \frac{1}{p-(-2)}$$

$$= \frac{1}{p+2}$$

Chap. 10 Using The Table To Find Laplace Transforms

R* find the Laplace transform of $b(x) = e^{-3x}$

$$b(x) = e^{-3x}$$

$$L[e^{kx}] = \frac{1}{p-k}$$

$$L[e^{-3x}] = \frac{1}{p-(-3)}$$

$$= \frac{1}{p+3}$$

4* find the Laplace transform of $b(x) = e^{-4x}$

$$b(x) = e^{-4x}$$

$$L[e^{kx}] = \frac{1}{p-k}$$

$$L[e^{-4x}] = \frac{1}{p-(-4)}$$

$$= \frac{1}{p+4}$$

Chapt 10 Using The Table To Find Laplace Transforms

ex | find the Laplace transform of $f(x) = e^{\frac{x}{2}}$

$$f(x) = e^{\frac{x}{2}}$$

$$L[e^{kx}] = \frac{1}{p - k}$$

$$L[e^{\frac{x}{2}}] = \frac{1}{p - \frac{1}{2}}$$

$$= \frac{2}{2}\left(\frac{1}{p - \frac{1}{2}}\right)$$

$$= \frac{2}{2p - 1}$$

＊ | find the Laplace transform of $f(x) = e^{-\frac{x}{2}}$

$$f(x) = e^{-\frac{x}{2}}$$

$$L[e^{kx}] = \frac{1}{p - k}$$

$$L[e^{-\frac{x}{2}}] = \frac{1}{p - (-\frac{1}{2})}$$

$$= \frac{1}{p + \frac{1}{2}}$$

$$= \frac{2}{2}\left(\frac{1}{p + \frac{1}{2}}\right)$$

$$= \frac{2}{2p + 1}$$

(116)

Ceration Using The Table To Find Laplace Transform trigonometric function

Q: find the Laplace transform of $f(x) = \sin x$

$$f(x) = \sin x$$

$$L[\sin Kx] = \frac{K}{p^2 + K^2}$$

$$L[\sin x] = \frac{1}{p^2 + 1^2}$$

$$= \frac{1}{p^2 + 1}$$

Q: find the Laplace transform of $f(x) = \sin 2x$

$$f(x) = \sin 2x$$

$$L[\sin Kx] = \frac{K}{p^2 + K^2}$$

$$L[\sin 2x] = \frac{2}{p^2 + 2^2}$$

$$= \frac{2}{p^2 + 4}$$

chapter Using The Table To Find Laplace Transform

ex find the Laplace Transform of $f(x) = \sin 3x$

$f(x) = \sin 3x$

$L\left[\sin kx\right] = \dfrac{k}{p^2 + k^2}$

$L\left[\sin 3x\right] = \dfrac{3}{p^2 + 3^2}$

$= \dfrac{3}{p^2 + 9}$

ex find the Laplace Transform of $f(x) = \sin 4x$

$f(x) = \sin 4x$

$L\left[\sin kx\right] = \dfrac{k}{p^2 + k^2}$

$L\left[\sin 4x\right] = \dfrac{4}{p^2 + 4^2}$

$= \dfrac{4}{p^2 + 16}$

Chapt 10 Using The Table To Find Laplace Transform

2x find the Laplace transform of $f(x) = \cos x$

$f(x) = \cos x$

$L[\cos Kx] = \dfrac{p}{p^2 + K^2}$

$L[\cos x] = \dfrac{p}{p^2 + 1^2}$

$\qquad = \dfrac{p}{p^2 + 1}$

2x find the Laplace transform of $f(x) = \cos 2x$

$f(x) = \cos 2x$

$L[\cos Kx] = \dfrac{p}{p^2 + K^2}$

$L[\cos 2x] = \dfrac{p}{p^2 + 2^2}$

$\qquad = \dfrac{p}{p^2 + 4}$

chapter 10 Using The Table To Find Laplace Transform

xx find the Laplace transform of $f(x) = \cos 3x$

$f(x) = \cos 3x$

$L[\cos kx] = \dfrac{p}{p^2 + k^2}$

$L[\cos 3x] = \dfrac{p}{p^2 + 3^2}$

$= \dfrac{p}{p^2 + 9}$

xx find the Laplace transform of $f(x) = \cos 4x$

$f(x) = \cos 4x$

$L[\cos kx] = \dfrac{p}{p^2 + k^2}$

$L[\cos 4x] = \dfrac{p}{p^2 + 4^2}$

$= \dfrac{p}{p^2 + 16}$

(120)

Chapter 10 Using The Table To Find Laplace Transform hyperbolic functions

&x find the Laplace transform of $f(x) = \sinh x$

$f(x) = \sinh x$

$$L[\sinh Kx] = \frac{K}{p^2 - K^2}$$

$$L[\sinh x] = \frac{1}{p^2 - 1^2}$$

$$= \frac{1}{p^2 - 1}$$

&x find the Laplace transform of $f(x) = \sinh 2x$

$f(x) = \sinh 2x$

$$L[\sinh Kx] = \frac{K}{p^2 - K^2}$$

$$L[\sinh 2x] = \frac{2}{p^2 - 2^2}$$

$$= \frac{2}{p^2 - 4}$$

<u>Chapter 10 Using The Table To Find Laplace Transform</u>

ex find the Laplace transform of $f(x) = \cosh x$

$f(x) = \cosh x$

$L[\cosh kx] = \dfrac{p}{p^2 - k^2}$

$L[\cosh x] = \dfrac{p}{p^2 - 1^2}$

$\qquad = \dfrac{p}{p^2 - 1}$

ex find the Laplace transform of $f(x) = \cosh 2x$

$f(x) = \cosh 2x$

$L[\cosh kx] = \dfrac{p}{p^2 - k^2}$

$L[\cosh 2x] = \dfrac{p}{p^2 - 2^2}$

$\qquad = \dfrac{p}{p^2 - 4}$

Chapter 10 Using The Table To Find Laplace Transforms

① using the definition of hyperbolic functions

$$\sinh x = \frac{e^x - e^{-x}}{2}$$

$$\cosh x = \frac{e^x + e^{-x}}{2}$$

demonstrate

$$L[\sinh x] = L\left[\frac{e^x - e^{-x}}{2}\right]$$

$$L[\cosh x] = L\left[\frac{e^x + e^{-x}}{2}\right]$$

(123)

(1 cont) Unit 10 Using The Table To Find Laplce Transform

$$L[\sinh kx] = \frac{k}{p^2 - k^2}$$

$$L[\sinh x] = \frac{1}{p^2 - 1^2}$$

$$L[\sinh x] = \frac{1}{p^2 - 1} \qquad \checkmark$$

$$L[\cosh kx] = \frac{p}{p^2 - k^2}$$

$$L[\cosh x] = \frac{p}{p^2 - 1^2}$$

$$L[\cosh x] = \frac{p}{p^2 - 1} \qquad \checkmark$$

(124)

(Cont) $L \left[\dfrac{e^x - e^{-x}}{2} \right]$

$= L \left[\dfrac{1}{2} (e^x - e^{-x}) \right]$

$= \dfrac{1}{2} L \left[e^x - e^{-x} \right]$

$= \dfrac{1}{2} \left(\dfrac{1}{p-1} - \dfrac{1}{p+1} \right)$

$= \dfrac{1}{2} \left(\dfrac{1}{p-1} \cdot \dfrac{p+1}{p+1} - \dfrac{1}{p+1} \cdot \dfrac{p-1}{p-1} \right)$

$= \dfrac{1}{2} \left[\dfrac{p+1 - p+1}{(p-1)(p+1)} \right]$

$= \dfrac{1}{2} \left[\dfrac{2}{(p-1)(p+1)} \right]$

$= \dfrac{1}{(p-1)(p+1)}$

$= \dfrac{1}{p^2 - 1}$

ID   Prob

(1 cont) $L\left[\dfrac{e^x + e^{-x}}{2}\right]$

$= L\left[\dfrac{1}{2}\left(e^x + e^{-x}\right)\right]$

$= \dfrac{1}{2} L\left[e^x + e^{-x}\right]$

$= \dfrac{1}{2}\left(\dfrac{1}{p-1} + \dfrac{1}{p+1}\right)$

$= \dfrac{1}{2}\left(\dfrac{1}{p-1}\cdot\dfrac{p+1}{p+1} + \dfrac{1}{p+1}\cdot\dfrac{p-1}{p-1}\right)$

$= \dfrac{1}{2}\left[\dfrac{p+1 + p-1}{(p-1)(p+1)}\right]$

$= \dfrac{1}{2}\left[\dfrac{2p}{(p-1)(p+1)}\right]$

$= \dfrac{p}{(p-1)(p+1)}$

$= \dfrac{p}{p^2 - 1}$

chapt 11   The Laplace Transformation to Linear
Linear Transformations

① an operator T and its corresponding transformation
is linear if

$$T[a \, f(x)] = a \, T[f(x)] \qquad \text{and}$$

$$T[f(x) + g(x)] = T[f(x)] + T[g(x)]$$

② from these two properties we may derive
a third property

$$T[a \, f(x) + b \, g(x)] = T[a \, f(x)] + T[b \, g(x)]$$
$$= a \, T[f(x)] + b \, T[g(x)]$$

③ we may show that the operations of
differentiation and
integration
are linear

(127)

Unit II The Laplace Transformation to Laws
Differentiation to A Linear Operation

① we must show
ⓐ $D[a\ f(x)] = a\ D[f(x)]$
ⓑ $D[f(x) + g(x)] = D[f(x)] + D[g(x)]$

ⓐ $\dfrac{d}{dx}[a\ f(x)] = a\ \dfrac{d}{dx} f(x)$

ⓑ $\dfrac{d}{dx}[f(x) + g(x)] = \dfrac{d}{dx} f(x) + \dfrac{d}{dx} g(x)$

note

the above two properties are in
the rules of differentiation

(128)

Chapt 11 The Laplace Transformation to Linear
Integration is A Linear Operation

① we must show

① $I[a\ f(x)] = a\ I[f(x)]$

② $I[f(x) + g(x)] = I[f(x)] + I[g(x)]$

① $\int a\ f(x)\ dx = a \int f(x)\ dx$

② $\int [f(x) + g(x)]\ dx = \int f(x)\ dx + \int g(x)\ dx$

note
the above two properties are in
the rules of integration

(129)

Chapt 11 The Laplace Transformation Is Linear

The Laplace Transformation Is Linear

① we must show

ⓐ $L[a f(x)] = a L[f(x)]$

ⓑ $L[f(x) + g(x)] = L[f(x)] + L[g(x)]$

ⓐ $L[a f(x)] = \int_0^{+\infty} a f(x) e^{-px} \, dx$

$= a \int_0^{+\infty} f(x) e^{-px} \, dx$

$= a \ L[f(x)]$

ⓑ $L[f(x) + g(x)] = \int_0^{+\infty} [f(x) + g(x)] e^{-px} \, dx$

$= \int_0^{+\infty} [f(x) e^{-px} + g(x) e^{-px}] \, dx$

$= \int_0^{+\infty} f(x) e^{-px} \, dx + \int_0^{+\infty} g(x) e^{-px} \, dx$

$= L[f(x)] + L[g(x)]$

② the following examples illustrate
the linearity of the Laplace Transformation

Chapt 11 The Laplace Transformation for Linear (130)

xx find the Laplace transform of
$f(x) = x^3 + x^2$

$L[x^3 + x^2]$

$= L[x^3] + L[x^2]$

$= \dfrac{3!}{p^{3+1}} + \dfrac{2!}{p^{2+1}}$

$= \dfrac{3 \cdot 2 \cdot 1}{p^4} + \dfrac{2 \cdot 1}{p^3}$

$= \dfrac{6}{p^4} + \dfrac{2}{p^3}$

xx find the Laplace transform of
$f(x) = 5x^4 + 3x$

$L[5x^5 + 3x]$

$= L[5x^5] + L[3x]$

$= 5 L[x^5] + 3 L[x]$

$= 5 \cdot \dfrac{5!}{p^{5+1}} + 3 \cdot \dfrac{1!}{p^{1+1}}$

$= 5\left(\dfrac{5 \cdot 4 \cdot 3 \cdot 2 \cdot 1}{p^6}\right) + 3\left(\dfrac{1}{p^2}\right)$

$= \dfrac{600}{p^6} + \dfrac{3}{p^2}$

chapt 11 The Laplace Transformation to Linear

ex   find the Laplace transform of

$f(x) = 2x^3 - 3x^2 + 7$

$L[2x^3 - 3x^2 + 7]$

$= L[2x^3] - L[3x^2] + L[7]$

$= 2L[x^3] - 3L[x^2] + L[7 \cdot 1]$

$= 2L[x^3] - 3L[x^2] + 7L[1]$

$= 2 \cdot \dfrac{3!}{p^{3+1}} - 3 \cdot \dfrac{2!}{p^{2+1}} + 7 \cdot \dfrac{1}{p}$

$= 2\left(\dfrac{3 \cdot 2 \cdot 1}{p^4}\right) - 3\left(\dfrac{2 \cdot 1}{p^3}\right) + \dfrac{7}{p}$

$= \dfrac{12}{p^4} - \dfrac{6}{p^3} + \dfrac{7}{p}$

note

$2x^3 - 3x^2 + 7$

$= 2x^3 + (-3)x^2 + 7$

therefore

$L[2x^3 - 3x^2 + 7]$

$= L[2x^3 + (-3)x^2 + 7]$

(132)

Chapt 11 The Laplace Transformation for Linear

also:

$$L[2x^3 + (-3)x^2 + 7]$$

$$= L[2x^3] + L[(-3)x^2] + L[7]$$

$$= 2\,L[x^3] + (-3)\,L[x^2] + L[7 \cdot 1]$$

$$= 2\,\frac{3!}{p^{3+1}} + (-3)\,\frac{2!}{p^{2+1}} + 7\,L[1]$$

$$= 2\left(\frac{3 \cdot 2 \cdot 1}{p^4}\right) + (-3)\left(\frac{2 \cdot 1}{p^3}\right) + 7 \cdot \frac{1}{p}$$

$$= \frac{12}{p^4} - \frac{6}{p^3} + \frac{7}{p}$$

i

chapt 11 The Laplace Transformation to Linear

ex find the Laplace transform of

$$f(x) = x^2 - e^{3x} + \sin 5x$$

$$L[x^2 - e^{3x} + \sin 5x]$$

$$= L[x^2] - L[e^{3x}] + L[\sin 5x]$$

$$= \frac{2!}{p^{2+1}} - \frac{1}{p-3} + \frac{5}{p^2 + 5^2}$$

$$= \frac{2 \cdot 1}{p^3} - \frac{1}{p-3} + \frac{5}{p^2 + 25}$$

$$= \frac{2}{p^3} - \frac{1}{p-3} + \frac{5}{p^2 + 25}$$

ex find the Laplace transform of

$$f(x) = 5 \sin 3x - 2 \cos 4x + 7 e^{-3x}$$

$$L[5 \sin 3x - 2 \cos 4x + 7 e^{-3x}]$$

$$= L[5 \sin 3x] - L[2 \cos 4x] + L[7 e^{-3x}]$$

$$= 5 L[\sin 3x] - 2 L[\cos 4x] + 7 L[e^{-3x}]$$

$$= 5 \cdot \frac{3}{p^2 + 3^2} - 2 \cdot \frac{p}{p^2 + 4^2} + 7 \cdot \frac{1}{p+3}$$

$$= \frac{15}{p^2 + 9} - \frac{2p}{p^2 + 16} + \frac{7}{p+3}$$

Chapt 11 The Laplace Transformation Is Linear

The Importance Of The Laplace Transformation Being Linear

① the linearity property of the Laplace transformation allows us to find the Laplace transform of a tremendous variety of functions involving sums and differences of functions multiplied by constants

(135)

Chapt 12 The Multiplication Formula
The Multiplication Formula

① the multiplication formula is given by

$$L[e^{kx} f(x)] = F(p-k)$$

where

$$L[f(x)] = F(p)$$

② this formula allows us to find the Laplace transform of the product involving e^{kx} and another function $f(x)$ which has a Laplace transform

③ in using the above formula we use two steps-
① find the Laplace transform of $f(x)$ and
② substitute $p-k$ for p

④ the following examples illustrate this technique

(136)

chpt 12 The multiplication Formula

ex find the Laplace transform of the function
$g(x) = e^{2x} x$

$$g(x) = e^{2x} x$$

$$L[e^{kx} b(x)] = F(p-k)$$

where

$$L[b(x)] = F(p)$$

$$L[x'] = \frac{1!}{p^{1+1}}$$

$$= \frac{1}{p^2}$$

$$L[e^{2x} x] = \frac{1}{(p-2)^2}$$

chapter 12 The Multiplication Formula

ex find the Laplace transform of the function

$g(x) = e^{2x} x^2$

$g(x) = e^{2x} x^2$

$L[e^{kx} g(x)] = F(p-k)$

where

$L[g(x)] = F(p)$

$L[x^2] = \dfrac{2!}{p^{2+1}}$

$= \dfrac{2}{p^3}$

$L[e^{2x} x^2] = \dfrac{2}{(p-2)^3}$

Chapt 12 The Multiplication Formula

ex   find the Laplace transform of the function
$g(x) = e^{2x} x^3$

$$g(x) = e^{2x} x^3$$

$$L[e^{kx} g(x)] = F(p - k)$$

where

$$L[g(x)] = F(p)$$

$$L[x^3] = \frac{3!}{p^{3+1}}$$

$$= \frac{3 \cdot 2 \cdot 1}{p^4}$$

$$= \frac{6}{p^4}$$

$$L[e^{2x} x^3] = \frac{6}{(p-2)^4}$$

chapt 12 The Multiplication Formula

Ex find the Laplace transform of the function

$$g(x) = e^{-3x} x^2$$

$$g(x) = e^{-3x} x^2$$

$$L[e^{kx} g(x)] = F(p-k)$$

where

$$L[g(x)] = F(p)$$

$$L[x^2] = \frac{2!}{p^{2+1}}$$

$$= \frac{2 \cdot 1}{p^3}$$

$$= \frac{2}{p^3}$$

$$L[e^{-3x} x^2] = \frac{2}{[p-(-3)]^3}$$

$$= \frac{2}{(p+3)^3}$$

)

$\sqrt{}$ | Chapter 12 The Multiplication Formula

find the Laplace transform of the function
$g(x) = e^{5x} \sin 2x$

$g(x) = e^{5x} \sin 2x$

$L[e^{kx} f(x)] = F(p-k)$

where

$L[f(x)] = F(p)$

$L[\sin 2x] = \dfrac{2}{p^2 + 2^2}$

$\qquad\qquad\quad = \dfrac{2}{p^2 + 4}$

$L[e^{5x} \sin 2x] = \dfrac{2}{(p-5)^2 + 4}$

$\qquad\qquad\quad = \dfrac{2}{p^2 - 10p + 25 + 4}$

$\qquad\qquad\quad = \dfrac{2}{p^2 - 10p + 29}$

Chapt 12 The Multiplication Formula (141)

ex find the Laplace transform of the function
$g(x) = e^{-5x} \sin 2x$

$g(x) = e^{-5x} \sin 2x$

$L[e^{kx} g(x)] = F(p-k)$
where
$L[g(x)] = F(p)$

$L[\sin 2x] = \dfrac{2}{p^2 + 2^2}$

$\qquad = \dfrac{2}{p^2 + 4}$

$L[e^{-5x} \sin 2x] = \dfrac{2}{(p+5)^2 + 4}$

$\qquad = \dfrac{2}{p^2 + 10p + 25 + 4}$

$\qquad = \dfrac{2}{p^2 + 10p + 29}$

$\overline{\text{chapt 12 The Multiplication Formula}}$

$\boxed{142}$

xx find the Laplace transform of the function

$g(x) = x^{3x} \cos x$

$g(x) = e^{3x} \cos x$

$L\left[e^{kx} f(x)\right] = F(p-k)$

where

$L\left[f(x)\right] = F(p)$

$L\left[\cos x\right] = \dfrac{p}{p^2 + 1^2}$

$= \dfrac{p}{p^2 + 1}$

$L\left[e^{3x} \cos x\right] = \dfrac{p-3}{(p-3)^2 + 1}$

$= \dfrac{p-3}{p^2 - 6p + 9 + 1}$

$= \dfrac{p-3}{p^2 - 6p + 10}$

Chapter 12 The Multiplication Formula

ex find the Laplace transform of the function

$g(x) = e^{-3x} \cos x$

$g(x) = e^{-3x} \cos x$

$L[e^{kx} g(x)] = F(p-k)$

where

$L[g(x)] = F(p)$

$L[\cos x] = \dfrac{p}{p^2 + 1^2}$

$ = \dfrac{p}{p^2 + 1}$

$L[e^{-3x} \cos x] = \dfrac{p+3}{(p+3)^2 + 1}$

$\phantom{L[e^{-3x} \cos x]} = \dfrac{p+3}{p^2 + 6p + 9 + 1}$

$\phantom{L[e^{-3x} \cos x]} = \dfrac{p+3}{p^2 + 6p + 10}$

chapt 12 The Multiplication Formula

4+ find the Laplace transform of the function
$g(x) = e^{4x} \sin(-5x)$

$g(x) = e^{4x} \sin(-5x)$ $\sin(-A) = -\sin A$
$g(x) = e^{4x} (-\sin 5x)$
$g(x) = -e^{4x} \sin 5x$

$L[e^{kx} f(x)] = F(p-k)$
where
$L[f(x)] = F(p)$

$L[\sin 5x] = \dfrac{5}{p^2 + 5^2}$

$\qquad\qquad = \dfrac{5}{p^2 + 25}$

$L[e^{4x} \sin 5x] = \dfrac{5}{(p-4)^2 + 25}$

$\qquad\qquad = \dfrac{5}{p^2 - 8p + 16 + 25}$

$\qquad\qquad = \dfrac{5}{p^2 - 8p + 41}$

$= \begin{aligned}[t] & L[-e^{4x} \sin 5x] \\ & - L[e^{4x} \sin 5x] \end{aligned}$

$= - \dfrac{5}{p^2 - 8p + 41}$

chapt 12 The Multiplication Formula

Find the Laplace transform of the function

$$g(x) = e^{-3x} \cos(-2x)$$

$$g(x) = e^{-3x} \cos(-2x) \qquad \cos(-A) = \cos A$$
$$g(x) = e^{-3x} \cos 2x$$

$$L\left[e^{kx} g(x)\right] = F(p-k)$$

where

$$L\left[g(x)\right] = F(p)$$

$$L\left[\cos 2x\right] = \frac{p}{p^2 + 2^2}$$

$$= \frac{p}{p^2 + 4}$$

$$L\left[e^{-3x} \cos 2x\right] = \frac{p+3}{(p+3)^2 + 4}$$

$$= \frac{p+3}{p^2 + 6p + 9 + 4}$$

$$= \frac{p+3}{p^2 + 6p + 13}$$

(146)

Chapt 12 The Multiplication Formula

Using The Multiplication Formula
To find The Laplace Transforms
OF General Products

① we may use the multiplication formula

$$L\left[e^{kx} g(x)\right] = F(p-k)$$

where

$$L\left[g(x)\right] = F(p)$$

to find the Laplace transforms of the following general products

$$g(x) = e^{ax} x^n \qquad\qquad n = 0, 1, 2, 3, \ldots$$

$$g(x) = e^{ax} \sin bx$$

$$g(x) = e^{ax} \cos bx$$

$$g(x) = e^{ax} \sinh bx$$

$$g(x) = e^{ax} \cosh bx$$

where a and b are constants

Chapt 12 The Multiplication Formula

$$g(x) = e^{ax} x^n$$

$$L[e^{ax} f(x)] = F(p-k)$$

where

$$L[f(x)] = F(p)$$

$$L[x^n] = \frac{n!}{p^{n+1}} \qquad n = 0, 1, 2, 3, \ldots$$

$$L[e^{ax} x^n] = \frac{n!}{(p-a)^{n+1}} \qquad n = 0, 1, 2, 3, \ldots$$

Chapt 12 The Multiplication Formula

$$g(x) = e^{ax} \sin bx$$
$$L\left[e^{kx} f(x)\right] = F(p-k)$$
where
$$L\left[f(x)\right] = F(p)$$

$$L\left[\sin bx\right] = \frac{b}{p^2 + b^2}$$

$$L\left[e^{ax} \sin bx\right] = \frac{b}{(p-a)^2 + b^2}$$

$$g(x) = e^{ax} \cos bx$$
$$L\left[e^{kx} f(x)\right] = F(p-k)$$
where
$$L\left[f(x)\right] = F(p)$$

$$L\left[\cos bx\right] = \frac{p}{p^2 + b^2}$$

$$L\left[e^{ax} \cos bx\right] = \frac{p-a}{(p-a)^2 + b^2}$$

Chapt 12 The Multiplication Formula

$$g(x) = e^{ax} \sinh bx$$
$$L[e^{kx} f(x)] = F(p-k)$$

where

$$L[f(x)] = F(p)$$

$$L[\sinh bx] = \frac{b}{p^2 - b^2}$$

$$L[e^{ax} \sinh bx] = \frac{b}{(p-a)^2 - b^2}$$

$$g(x) = e^{ax} \cosh bx$$
$$L[e^{kx} f(x)] = F(p-k)$$

where

$$L[f(x)] = F(p)$$

$$L[\cosh bx] = \frac{p}{p^2 - b^2}$$

$$L[e^{ax} \cosh bx] = \frac{p-a}{(p-a)^2 - b^2}$$

(149)

Chapt 12 The Multiplication Formula

② In summary we may write

$$L\left[e^{ax}\,x^n\right] = \frac{n!}{(p-a)^{n+1}} \qquad n = 0, 1, 2, 3, \ldots$$

$$L\left[e^{ax}\,\sin bx\right] = \frac{b}{(p-a)^2 + b^2}$$

$$L\left[e^{ax}\,\cos bx\right] = \frac{p-a}{(p-a)^2 + b^2}$$

$$L\left[e^{ax}\,\sinh bx\right] = \frac{b}{(p-a)^2 - b^2}$$

$$L\left[e^{ax}\,\cosh bx\right] = \frac{p-a}{(p-a)^2 - b^2}$$

(150)

Chpt 12 The Multiplication Formula

① prove the multiplication formula

$$L[e^{kx} f(x)] = F(p-k)$$

where

$$L[f(x)] = F(p)$$

__Proof__

$$L[f(x)] = \int_0^{+\infty} f(x) \, e^{-px} \, dx = F(p)$$

$$L[e^{kx} f(x)] = \int_0^{+\infty} [e^{kx} f(x)] \, e^{-px} \, dx$$

$$L[e^{kx} f(x)] = \int_0^{+\infty} e^{kx} f(x) \, e^{-px} \, dx$$

$$L[e^{kx} f(x)] = \int_0^{+\infty} f(x) \, e^{kx} \, e^{-px} \, dx$$

$$L[e^{kx} f(x)] = \int_0^{+\infty} f(x) \, e^{kx-px} \, dx$$

$$L[e^{kx} f(x)] = \int_0^{+\infty} f(x) \, e^{(k-p)x} \, dx$$

$$L[e^{kx} f(x)] = \int_0^{+\infty} f(x) \, e^{-(p-k)x} \, dx$$

let $p^* = p - k$

$$L[e^{kx} f(x)] = \int_0^{+\infty} f(x) \, e^{-p^* x} \, dx$$

$$L[e^{kx} f(x)] = F(p^*)$$
$$L[e^{kx} f(x)] = F(p-k)$$

Chpt 12 The Multiplication Formula

also ⓐ prove the multiplication formula

$$L\left[e^{kx} f(x)\right] = F(p-k)$$

where

$$L\left[f(x)\right] = F(p)$$

proof $L\left[f(x)\right] = \int_{0}^{+\infty} f(x) e^{-px} dx = F(p)$

$$F(p) = \int_{0}^{+\infty} f(x) e^{-px} dx$$

$$F(p-k) = \int_{0}^{+\infty} f(x) e^{-(p-k)x} dx$$

$$F(p-k) = \int_{0}^{+\infty} f(x) e^{-px+kx} dx$$

$$F(p-k) = \int_{0}^{+\infty} f(x) e^{-px} e^{kx} dx$$

$$F(p-k) = \int_{0}^{+\infty} e^{kx} f(x) e^{-px} dx$$

$$F(p-k) = \int_{0}^{+\infty} \left[e^{kx} f(x)\right] e^{-px} dx$$

let $g(x) = e^{kx} f(x)$

$$F(p-k) = \int_{0}^{+\infty} g(x) e^{-px} dx$$

$$\int_{0}^{+\infty} g(x) e^{-px} dx = F(p-k)$$

where

$$g(x) = e^{kx} f(x)$$

Chapt 12 The Multiplication Formula

(2 cont) $L[g(x)] = F(\rho - \kappa)$

$$L[e^{\kappa x} f(x)] = F(\rho - \kappa)$$

where

Proof $\quad L[f(x)] = F(\rho)$

Chapt 13 The Laplace Transformation of A Derivative

The Laplace Transformation Of A Derivative

① we may use the definition of the Laplace transformation of a function

$$y = f(x)$$
$$L[f(x)] = \int_0^{+\infty} f(x) e^{-px} dx$$

to find the Laplace transform of the function's first derivative

$$\frac{dy}{dx} = f'(x)$$

and

the function's second derivative

$$\frac{d^2y}{dx^2} = f''(x)$$

and

higher derivatives of the function

② the results we obtain may be stated as the following formulas

$$L[y'] = p\, L[y] - y(0)$$
$$L[y''] = p^2\, L[y] - p\, y(0) - y'(0)$$
$$L[y'''] = p^3\, L[y] - p^2\, y(0) - p\, y'(0) - y''(0)$$
$$L[y''''] = p^4\, L[y] - p^3\, y(0) - p^2\, y'(0) - p\, y''(0) - y'''(0)$$

etc

③ the following examples illustrate the relationships in each of the first two formulas

Chapt 13 The Laplace Transformation Of A Derivative

ex: consider the function $y = 1$

verify

$$L[y'] = p L[y] - y(0)$$

$$y = 1 \qquad y(x) = 1$$
$$y' = 0 \qquad y(0) = 1$$

$$L[0] = p L[1] - 1$$
$$= p \left(\frac{1}{p} \right) - 1$$
$$= 1 - 1$$
$$= 0$$

$$L[0] = \frac{0}{p}$$
$$= 0$$

$\underline{\text{Chapt 13 The Laplace Transformation Of A Derivative}}$

consider the function $y = x$

verify

$L[y'] = p \, L[y] - y(0)$

$\begin{aligned} y &= x \\ y' &= 1 \end{aligned}$ $\qquad$ $\begin{aligned} y(x) &= x \\ y(0) &= 0 \end{aligned}$

$$L[1] = p \, L[x] - 0$$
$$= p \left(\frac{1!}{p^{1+1}} \right) - 0$$
$$= p \left(\frac{1}{p^2} \right)$$
$$= \frac{1}{p}$$

$$L[1] = \frac{1}{p}$$

chapt 13 The Laplace Transformation Of A Derivative

Ex consider the function $y = x^2$

verify

$$L[y'] = p\, L[y] - Y(0)$$

$$y = x^2 \qquad\qquad y(x) = x^2$$
$$y' = 2x \qquad\qquad y(0) = 0^2 = 0$$

$$L[2x] = p\, L[x^2] - 0$$
$$= p\left(\frac{2!}{p^{2+1}}\right) - 0$$
$$= p\left(\frac{2\cdot 1}{p^3}\right)$$
$$= p\left(\frac{2}{p^3}\right)$$
$$= \frac{2}{p^2}$$

$$= L[2x]$$
$$= 2\,L[x']$$
$$= 2\left(\frac{1!}{p^{1+1}}\right)$$
$$= 2\,\frac{1}{p^2}$$
$$= \frac{2}{p^2}$$

Chapt 13 The Laplace Transformation Of A Derivative

consider the function $y = x^3$

$L[y'] = p L[y] - y(0)$

$y_1 = x^3$ $\qquad\qquad$ $y(x) = x^3$

$y' = 3x^2$ $\qquad\qquad$ $y(0) = 0^3 = 0$

$L[3x^2] = p L[x^3] - 0$

$\qquad\qquad = p \left(\dfrac{3!}{p^{3+1}} \right) - 0$

$\qquad\qquad = p \left(\dfrac{6}{p^4} \right)$

$\qquad\qquad = \dfrac{6}{p^3}$

$\qquad L[3x^2]$

$= 3 L[x^2]$

$= 3 \left(\dfrac{2!}{p^{2+1}} \right)$

$= 3 \left(\dfrac{2}{p^3} \right)$

$= \dfrac{6}{p^3}$

(a)(i) The Laplace Transform of A Derivative

ex   consider the function $y = x^3$

verify

$$L[y''] = p^2 L[y] - p\, y(0) - y'(0)$$

$$\begin{aligned} y &= x^3 \\ y' &= 3x^2 \\ y'' &= 6x \end{aligned} \qquad \begin{aligned} y(x) &= x^3 \\ y(0) &= 0^3 \\ y(0) &= 0 \end{aligned} \qquad \begin{aligned} y'(x) &= 3x^2 \\ y'(0) &= 3 \cdot 0^2 \\ y'(0) &= 0 \end{aligned}$$

$$\begin{aligned} L[6x] &= p^2 L[x^3] - p(0) - 0 \\ &= p^2 \left(\frac{3!}{p^{3+1}} \right) - 0 - 0 \\ &= p^2 \left(\frac{6}{p^4} \right) \\ &= \frac{6}{p^2} \end{aligned}$$

$$\begin{aligned} &\quad L[6x] \\ &= 6\, L[x'] \\ &= 6 \left(\frac{1!}{p^{1+1}} \right) \\ &= 6 \left(\frac{1}{p^2} \right) \\ &= \frac{6}{p^2} \end{aligned}$$

Chapt 13 The Laplace Transformation Of A Derivative

The Importance Of the Laplace

Derivative Formulas

① the importance of the formulas

$$L[y'] = p \, L[y] \, b - y(0) \qquad \text{and}$$

$$L[y''] = p^2 \, L[y] - p \, y(0) - y'(0)$$

is that they allow us to solve

initial value problems $(I.V.P's)$

involving differential equations

② note

the initial conditions are contained

in each formula —

$$y(0) \qquad \text{and} \qquad y'(0)$$

Chapt 13 The Laplace Transformation Of A Derivative

① derive the derivative formula

$$L[y'] = p \, L[y] - y(0)$$

proof

The
Derivative
Formula

let $y = f(x)$

also $y(x) = f(x)$

$$L[f(x)] = \int_0^{+\infty} f(x) \, e^{-px} \, dx$$

$$L\left[\frac{dy}{dx}\right] = \int_0^{+\infty} \frac{dy}{dx} \, e^{-px} \, dx$$

Chapt 13 The Laplace Transformation Of A Derivative

(cont) $\quad \int \dfrac{dy}{dx} \; e^{-px} \; dx$

$\int e^{-px} \; \dfrac{dy}{dx} \; dx$

proofs

The
Derivative
Formula

$\int u \; dv = uv - \int v \; du$

let $\quad u = e^{-px}$

$\dfrac{du}{dx} = e^{-px}(-p)$

$\dfrac{du}{dx} = -p \, e^{-px}$

$du = -p \, e^{-px} \; dx$

$dv = \dfrac{dy}{dx} \; dx$

$dv = dy$

$\int dv = \int dy$

$v = y$

$\int e^{-px} \; \dfrac{dy}{dx} \; dx$

$= e^{-px}(y) - \int y \, (-p \, e^{-px} \; dx)$

$= e^{-px} \, y + p \int y \, e^{-px} \; dx$

$= e^{-px} \, f(x) + p \int f(x) \, e^{-px} \; dx$

$\int^{on} \dfrac{dy}{dx} \; e^{-px} \; dx$

$= e^{-px} \, f(x) + p \int f(x) \, e^{-px} \; dx$

Chapt 13 The Laplace Transformation Of A Derivative

(cont)
$$\int_0^{+\infty} \frac{dy}{dx} \, e^{-px} \, dx$$

$$= e^{-px} \, f(x) \Big|_0^{+\infty} + p \int_0^{+\infty} f(x) \, e^{-px} \, dx$$

$$= \frac{f(x)}{e^{px}} \Big|_0^{+\infty} + p \, L[f(x)]$$

$*$
$$= 0 - \frac{f(0)}{e^0} + p \, L[f(x)]$$

$$= - \frac{f(0)}{1} + p \, L[f(x)]$$

$$= - f(0) + p \, L[f(x)]$$

$$= p \, L[f(x)] - f(0)$$

$$= p \, L[y] - y(0)$$

$*$ note

we require the following property of $f(x)$

$$\frac{f(x)}{e^{px}} \to 0 \qquad as \qquad x \to +\infty$$

(163)

Chapt 13 The Laplace Transformation Of A Derivative

② derive the derivative formula

$$L[y''] = p^2 L[y] - p\, y(0) - y'(0)$$

The
Derivative
Formula

$$L[y'] = p\, L[y] - y(0)$$

$$
\begin{aligned}
L[y''] &= p\, L[y'] - y'(0) \\
&= p\, \{ p\, L[y] - y(0) \} - y'(0) \\
&= p^2 L[y] - p\, y(0) - y'(0)
\end{aligned}
$$

note

consider the two formulas

$$L[y'] = p\, L[y] - y(0)$$
$$L[y''] = p\, L[y'] - y'(0)$$

the top formula is stating a relationship
between some function y and
the function's derivative y'
therefore
we may consider the function y' as
a function in its own right and
repeat this relationship as given
by the top formula which results in
the bottom formula

chapt 13 The Laplace Transform Of A Derivative

③ derive the derivative formula

$$L[y'''] = p^3 L[y] - p^2 y(0) - p y'(0) - y''(0)$$

proof

$$L[y'] = p L[y] - y(0)$$
$$L[y''] = p^2 L[y] - p y(0) - y'(0)$$

$$L[y'''] = p^2 L[y'] - p y'(0) - y''(0)$$
$$L[y'''] = p^2 \{ p L[y] - y(0) \} - p y'(0) - y''(0)$$
$$L[y'''] = p^3 L[y] - p^2 y(0) - p y'(0) - y''(0)$$

note
consider the two formulas
$$L[y''] = p^2 L[y] - p y(0) - y'(0)$$
$$L[y'''] = p^2 L[y'] - p y'(0) - y''(0)$$

the top formula is stating a relationship
between some function y and its
first two derivatives
therefore
we may consider the function y' as
a function in its own right and
represent the relationship as given by
the top formula which results in
the bottom formula

④ Unit 13 The Laplace Transform of A Derivative

derive the derivative formula

$$L[y''''] = p^4 L[y] - p^3 y(0) - p^2 y'(0) - p y''(0) - y'''(0)$$

Proof

$$L[y'] = p L[y] - y(0)$$

$$L[y''] = p^2 L[y] - p y(0) - y'(0)$$

$$L[y'''] = p^3 L[y] - p^2 y(0) - p y'(0) - y''(0)$$

$$L[y''''] = p^3 L[y'] - p^2 y'(0) - p y''(0) - y'''(0)$$

$$L[y''''] = p^3 \{ p L[y] - y(0) \} - p^2 y'(0) - p y''(0) - y'''(0)$$

$$L[y''''] = p^4 L[y] - p^3 y(0) - p^2 y'(0) - p y''(0) - y'''(0)$$

note

consider the two formulas

$$L[y'''] = p^3 L[y] - p^2 y(0) - p[y'(0)] - y''(0)$$

$$L[y''''] = p^3 L[y'] - p^2 y'(0) - p y''(0) - y'''(0)$$

the top formula is stating a relationship
between some function y and its
first three derivatives
therefore
we may consider the function y' as
a function in its own right and
repeat the relationship as given by
the top formula which results in the bottom formula

$$(166)$$

Chapt 14 The Inverse Laplace Transform

The Inverse Laplace Transform

① we also made use of our table to find inverse Laplace transforms

function $f(x)$ $\underset{L^{-1}}{\overset{L}{\rightleftharpoons}}$	Laplace Transform $F(p)$
0	0
1	$\dfrac{1}{p}$
K	$\dfrac{K}{p}$
x^n $n = 0, 1, 2, 3, \ldots$	$\dfrac{n!}{p^{n+1}}$
e^{Kx}	$\dfrac{1}{p - K}$
$\sin Kx$	$\dfrac{K}{p^2 + K^2}$
$\cos Kx$	$\dfrac{p}{p^2 + K^2}$
$\sinh Kx$	$\dfrac{K}{p^2 - K^2}$
$\cosh Kx$	$\dfrac{p}{p^2 - K^2}$

Chpt 14 The Inverse Laplace Transform

(2) ① to find the Laplace transform
we start with the function $f(x)$ and
find its corresponding function $F(p)$

② to find the inverse Laplace transform
we start with the function $F(p)$ and
find its corresponding function $f(x)$

(3) we may use arrow notation to indicate this

$$f(x) \xrightarrow{L} F(p)$$

$$f(x) \xleftarrow[L^{-1}]{} F(p)$$

or equivalently

$$f(x) \xrightarrow{L} F(p)$$

$$F(p) \xrightarrow[L^{-1}]{} f(x)$$

(4) the operator for the inverse Laplace transformation
is given the symbol L^{-1}

(5) note
what one operator does
the other undoes

(6) this is similar to the behavior of
the differentiation operator D and
the integration operator I
in calculus

(168)

Chapt 14 The Inverse Laplace Transform

⑦ using arrow notation

$$f(x) \xrightarrow{D} g(x)$$
$$f(x) \xleftarrow{I} g(x)$$

also

$$f(x) \xrightarrow{I} g(x)$$
$$f(x) \xleftarrow{D} g(x)$$

⑧ getting back to
the Laplace operator L   and
the inverse operator L⁻¹
and finding
Laplace transforms   and
inverse Laplace transforms

finding Laplace transforms
you are given $f(x)$ and must find $F(s)$

finding inverse Laplace transform
you are given $F(s)$ and must find $f(x)$

⑨ the following examples illustrate
the technique

(169)

Chpt 14 The Inverse Laplace Transform
constants

\# find the inverse Laplace transform of

$$F(p) = \frac{1}{p}$$

$$L^{-1}\left[\frac{K}{p}\right] = K$$

$$L^{-1}\left[\frac{1}{p}\right] = 1$$

or $f(x) = 1$

\# find the inverse Laplace transform of

$$F(p) = \frac{2}{p}$$

$$L^{-1}\left[\frac{K}{p}\right] = K$$

$$L^{-1}\left[\frac{2}{p}\right] = 2$$

or $f(x) = 2$

Chapt 14. The Inverse Laplace Transform

ex. Find the inverse Laplace transform of
$$F(p) = \frac{3}{p}$$

$$L^{-1}\left[\frac{k}{p}\right] = k$$

$$L^{-1}\left[\frac{3}{p}\right] = 3$$

or $f(x) = 3$

ex. Find the inverse Laplace transform of
$$F(p) = \frac{4}{p}$$

$$L^{-1}\left[\frac{k}{p}\right] = k$$

$$L^{-1}\left[\frac{4}{p}\right] = 4$$

or $f(x) = 4$

chapt 14 The Inverse Laplace Transform

1) find the inverse Laplace transform of
$$F(p) = - \frac{1}{p}$$

$$L^{-1}\left[\frac{K}{p}\right] = K$$

$$L^{-1}\left[\frac{-1}{p}\right] = -1$$

on $f(x) = -1$

2) find the inverse Laplace transform of
$$F(p) = - \frac{2}{p}$$

$$L^{-1}\left[\frac{K}{p}\right] = K$$

$$L^{-1}\left[\frac{-2}{p}\right] = -2$$

on $f(x) = -2$

$$(172)$$

ex | Chapt 14 The Inverse Laplace Transform
find the inverse Laplace transform of

$$F(p) = \frac{3}{7 p}$$

$$L^{-1}\left[\frac{K}{p}\right] = K$$

$$L^{-1}\left[\frac{3}{7 p}\right]$$

$$= L^{-1}\left[\frac{\frac{3}{7}}{p}\right]$$

$$= \frac{3}{7}$$

on $g(x) = \frac{3}{7}$

also

$$L^{-1}\left[\frac{3}{7 p}\right]$$

$$= L^{-1}\left[\frac{3}{7} \cdot \frac{1}{p}\right]$$

$$= \frac{3}{7} L^{-1}\left[\frac{1}{p}\right]$$

$$= \frac{3}{7} (1)$$

$$= \frac{3}{7}$$ on $g(x) = \frac{3}{7}$

note

in the next chapter
we will learn
the inverse Laplace operation
is linear

(173)

Chapt 14. The Inverse Laplace Transform

$*$ Find the inverse Laplace transform of

$$F(p) = \frac{7}{3p}$$

$$L^{-1}\left[\frac{K}{p}\right] = K$$

$$L^{-1}\left[\frac{7}{3p}\right]$$

$$= L^{-1}\left[\frac{\frac{7}{3}}{p}\right]$$

$$= \frac{7}{3}$$

or $f(x) = \frac{7}{3}$

also

$$L^{-1}\left[\frac{7}{3p}\right]$$

$$= L^{-1}\left[\frac{7}{3} \cdot \frac{1}{p}\right]$$

$$= \frac{7}{3} L^{-1}\left[\frac{1}{p}\right]$$

$$= \frac{7}{3} \cdot 1$$

$$= \frac{7}{3} \qquad or \quad f(x) = \frac{7}{3}$$

(174)

chapt 14 The Inverse Laplace Transform

polynomials

2) find the inverse Laplace transform of

$$F(\rho) = \frac{1}{\rho^2}$$

$$L^{-1}\left[\frac{n!}{\rho^{n+1}}\right] = X^n$$

$$L^{-1}\left[\frac{1}{\rho^2}\right]$$

$$= L^{-1}\left[\frac{1!}{\rho^{1+1}}\right]$$

$$n = 1$$

$$= X^1$$

$$= X$$

on $f(t) = X$

$\overline{X}$

ex

Chapt 14 The Inverse Laplace Transform

find the inverse Laplace transform of

$$F(p) = \frac{2}{p^3}$$

$$L^{-1}\left[\frac{n!}{p^{n+1}}\right] = x^n$$

$$L^{-1}\left[\frac{2}{p^3}\right]$$

$$= L^{-1}\left[\frac{2 \cdot 1}{p^{2+1}}\right]$$

$$= L^{-1}\left[\frac{2!}{p^{2+1}}\right]$$

$$n = 2$$

$$= x^2$$

or $\quad g(x) = x^2$

(176)

Chapt 14 The Inverse Laplace Transform

4. find the inverse Laplace transform of

$$F(\rho) = \frac{6}{\rho^4}$$

$$L^{-1}\left[\frac{n!}{\rho^{n+1}}\right] = x^n$$

$$L^{-1}\left[\frac{6}{\rho^4}\right]$$

$$= L^{-1}\left[\frac{3 \cdot 2 \cdot 1}{\rho^{3+1}}\right]$$

$$= L^{-1}\left[\frac{3!}{\rho^{3+1}}\right]$$

$$= x^3 \quad n = 3$$

$$\text{or} \quad f(x) = x^3$$

XII

dmtn·

(177)

Chapt 14 The Inverse Laplace Transform

Ex. find the inverse Laplace transform of

$F_{(\rho)} = \dfrac{7}{\rho^4}$

$L^{-1}\left[\dfrac{n!}{\rho^{n+1}}\right] = x^n$

$L^{-1}\left[\dfrac{7}{\rho^4}\right]$

$= L^{-1}\left[\dfrac{7}{\rho^{3+1}}\right]$

$= L^{-1}\left[\dfrac{7 \cdot \frac{3!}{3!}}{\rho^{3+1}}\right]$

$= \dfrac{7}{3!} L^{-1}\left[\dfrac{3!}{\rho^{3+1}}\right]$

$= \dfrac{7}{6} \cdot x^3$

$= \dfrac{7}{6} x^3$

or $f(x) = \dfrac{7}{6} x^3$

chapt 14 The inverse Laplace Transform

exponential functions

4↓ find the inverse Laplace transform of

$$F(\rho) = \frac{1}{\rho - 1}$$

$$L^{-1}\left[\frac{1}{\rho - K}\right] = e^{Kx}$$

$$L^{-1}\left[\frac{1}{\rho - 1}\right]$$

$$= e^x \qquad\qquad K = 1$$

on $f(x) = e^x$

4↓ find the inverse Laplace transform of

$$F(\rho) = \frac{1}{\rho - 2}$$

$$L^{-1}\left[\frac{1}{\rho - K}\right] = e^{Kx}$$

$$L^{-1}\left[\frac{1}{\rho - 2}\right]$$

$$= e^{2x} \qquad\qquad K = 2$$

on $f(x) = e^{2x}$

Chapt 14 The Inverse Laplace Transform

ex | find the inverse Laplace transform of

$$F(p) = \frac{1}{p-3}$$

$$L^{-1}\left[\frac{1}{p-K}\right] = e^{Kx}$$

$$L^{-1}\left[\frac{1}{p-3}\right]$$

$$= e^{3x} \qquad\qquad K = 3$$

on $\quad f(x) = e^{3x}$

ex | find the inverse Laplace transform of

$$F(p) = \frac{1}{p-4}$$

$$L^{-1}\left[\frac{1}{p-K}\right] = e^{Kx}$$

$$L^{-1}\left[\frac{1}{p-4}\right]$$

$$= e^{4x} \qquad\qquad K = 4$$

on $\quad f(x) = e^{4x}$

Chapt 14 The Inverse Laplace Transform

ex find the inverse Laplace transform of

$$F(p) = \frac{1}{p+1}$$

$$L^{-1}\left[\frac{1}{p-k}\right] = e^{kx}$$

$$L^{-1}\left[\frac{1}{p+1}\right]$$

$$= L^{-1}\left[\frac{1}{p-(-1)}\right]$$

$$= e^{-x} \qquad k = -1$$

or $f(x) = e^{-x}$

---

find the inverse Laplace transform of

$$F(p) = \frac{1}{p+2}$$

$$L^{-1}\left[\frac{1}{p-k}\right] = e^{kx}$$

$$L^{-1}\left[\frac{1}{p+2}\right]$$

$$= L^{-1}\left[\frac{1}{p-(-2)}\right] = e^{-2x} \qquad \text{or } f(x) = e^{-2x}$$

(181)

Chapt 14 The inverse Laplace Transform

ex Find the inverse Laplace transform of

$$F(p) = \frac{1}{p+3}$$

$$L^{-1}\left[\frac{1}{p-K}\right] = e^{Kx}$$

$$L^{-1}\left[\frac{1}{p+3}\right]$$

$$= L^{-1}\left[\frac{1}{p-(-3)}\right]$$

$$= e^{-3x} \qquad\qquad K = -3$$

on $f(x) = e^{-3x}$

ex Find the inverse Laplace transform of

$$F(p) = \frac{1}{p+4}$$

$$L^{-1}\left[\frac{1}{p-K}\right] = e^{Kx}$$

$$L^{-1}\left[\frac{1}{p+4}\right]$$

$$= L^{-1}\left[\frac{1}{p-(-4)}\right] = e^{-4x} \qquad on \qquad f(x) = e^{-4x}$$

(182)

Chapt 14 The Inverse Laplace Transform
Trigonometric functions

9+ find the inverse Laplace transform of

$$F(p) = \frac{1}{p^2 + 1}$$

$$L^{-1}\left[\frac{K}{p^2 + K^2}\right] = \sin Kx$$

$$L^{-1}\left[\frac{1}{p^2 + 1}\right]$$

$$= L^{-1}\left[\frac{1}{p^2 + 1^2}\right]$$

$$= \sin x \qquad\qquad K = 1$$

or $f(x) = \sin x$

Chapt 14 The Inverse Laplace Transform

2+ find the inverse Laplace transform of

$$F(p) = \frac{2}{p^2 + 4}$$

$$L^{-1}\left[\frac{k}{p^2 + k^2}\right] = \sin kx$$

$$L^{-1}\left[\frac{2}{p^2 + 4}\right]$$

$$= L^{-1}\left[\frac{2}{p^2 + 2^2}\right]$$

$$= \sin 2x \qquad\qquad k = 2$$

or $f(x) = \sin 2x$

chapt 14 The inverse Laplace Transform

Ex. find the inverse Laplace transform of

$$F(p) = \frac{3}{p^2 + 9}$$

$$L^{-1}\left[\frac{K}{p^2 + K^2}\right] = \sin KX$$

$$L^{-1}\left[\frac{3}{p^2 + 9}\right]$$

$$= L^{-1}\left[\frac{3}{p^2 + 3^2}\right]$$

$$= \sin 3X \qquad\qquad K = 3$$

or $\quad f(x) = \sin 3X$

(185)

Chapt 14 The Inverse Laplace Transform

ex find the inverse Laplace transform of

$$F(p) = \frac{4}{p^2 + 16}$$

$$L^{-1}\left[\frac{k}{p^2 + k^2}\right] = \sin kx$$

$$L^{-1}\left[\frac{4}{p^2 + 16}\right]$$

$$= L^{-1}\left[\frac{4}{p^2 + 4^2}\right]$$

$$= \sin 4x \qquad\qquad k = 4$$

$$\text{on} \quad f(x) = \sin 4x$$

Chapt 14 The inverse Laplace Transform ⑱⑥

&x find the inverse Laplace transform of

$$F(\rho) = \frac{\rho}{\rho^2 + 1}$$

$$L^{-1}\left[\frac{\rho}{\rho^2 + K^2}\right] = \cos Kx$$

$$= L^{-1}\left[\frac{\rho}{\rho^2 + 1}\right]$$

$$= L^{-1}\left[\frac{\rho}{\rho^2 + 1^2}\right]$$

$$= \cos x \qquad\qquad K = 1$$

so $\quad f(x) = \cos x$

Chapt 14 The Inverse Laplace Transform

Ex find the inverse Laplace transform of

$$F(p) = \frac{p}{p^2 + 4}$$

$$L^{-1}\left[\frac{p}{p^2 + K^2}\right] = \cos Kx$$

$$= L^{-1}\left[\frac{p}{p^2 + 4}\right]$$

$$= L^{-1}\left[\frac{p}{p^2 + 2^2}\right]$$

$$= \cos 2x \qquad\qquad K = 2$$

on $f(x) = \cos 2x$

Intro.

Chapt 14. The inverse Laplace Transform

ex find the inverse Laplace transform of

$$F(p) = \frac{p}{p^2 + 9}$$

$$L^{-1}\left[\frac{p}{p^2 + K^2}\right] = \cos KX$$

$$L^{-1}\left[\frac{p}{p^2 + 9}\right]$$

$$= L^{-1}\left[\frac{p}{p^2 + 3^2}\right]$$

$$= \cos 3X \qquad\qquad K = 3$$

or $\quad g(x) = \cos 3X$

XXIV

Intro

(189)

Chapt 14 The Inverse Laplace Transform

find the inverse Laplace transform if

$$F(p) = \frac{p}{p^2 + 16}$$

$$L^{-1}\left[\frac{p}{p^2 + K^2}\right] = \cos KX$$

$$L^{-1}\left[\frac{p}{p^2 + 16}\right]$$

$$= L^{-1}\left[\frac{p}{p^2 + 4^2}\right]$$

$$= \cos 4X \qquad\qquad K = 4$$

Or $\quad f(x) = \cos 4X$

(190)

chapt14 The inverse Laplace Transform
hyperbolic functions

4X find the inverse Laplace transform of

$$F(p) = \frac{1}{p^2 - 1}$$

$$L^{-1}\left[\frac{K}{p^2 - K^2}\right] = \sinh KX$$

$$L^{-1}\left[\frac{1}{p^2 - 1}\right]$$

$$= L^{-1}\left[\frac{1}{p^2 - 1^2}\right]$$

$$= \sinh x \qquad K = 1$$

or $g(x) = \sinh x$

Ex

(191)

Chapt 14 The ~~Inverse~~ Laplace Transform

find the inverse Laplace transform of

$$F(p) = \frac{4}{p^2 - 4}$$

$$L^{-1}\left[\frac{K}{p^2 - K^2}\right] = \sinh Kx$$

$$L^{-1}\left[\frac{2}{p^2 - 4}\right]$$

$$= L^{-1}\left[\frac{2}{p^2 - 2^2}\right]$$

$$= \sinh 2x \qquad\qquad K = 2$$

on $f(x) = \sinh 2x$

(192)

Chapt 14 The inverse Laplace Transform

Ex | find the inverse Laplace transform of

$$F(p) = \frac{p}{p^2 - 1}$$

$$L^{-1}\left[\frac{p}{p^2 - K^2}\right] = \cosh KX$$

$$L^{-1}\left[\frac{p}{p^2 - 1}\right]$$

$$= L^{-1}\left[\frac{p}{p^2 - 1^2}\right]$$

$$= \cosh X \qquad K = 1$$

or $\quad f(x) = \cosh X$

XXVIII

Intro

(193)

Chapt 14 The Inverse Laplace Transform

Find the inverse Laplace transform of

$$F(p) = \frac{p}{p^2 - 4}$$

$$L^{-1}\left[\frac{p}{p^2 - K^2}\right] = \cosh Kx$$

$$L^{-1}\left[\frac{p}{p^2 - 4}\right]$$

$$= L^{-1}\left[\frac{p}{p^2 - 2^2}\right]$$

$$= \cosh 2x \qquad\qquad K = 2$$

or $f(x) = \cosh 2x$

(194)

Chapt 14 The Inverse Laplace Transform
The Relationship Between The Exponential Function
And The Hyperbolic Functions

① the hyperbolic functions are defined
in terms of exponential functions

$$\sinh x = \frac{e^x - e^{-x}}{2}$$

$$\cosh x = \frac{e^x + e^{-x}}{2}$$

② this relationship reveals itself
in the denominators of the Laplace transforms

$$L[e^{kx}] = \frac{1}{p-k} \qquad ①$$

$$L[\sinh kx] = \frac{k}{p^2 - k^2} \qquad ②$$

$$L[\cosh kx] = \frac{p}{p^2 - k^2} \qquad ③$$

③ note
① the denominators of the Laplace transforms
of the hyperbolic functions may be factored
② furthermore using the method of partial fractions
we may place the Laplace transform ② or ③
as a sum or difference of form ①
③ the denominators of the Laplace transform
of the trigonometric functions may not
be factored because of the plus sign

(195)

Chapt 14 The Inverse Laplace Transform

$$L[\sin KX] = \frac{K}{P^2 + K^2}$$

$$L[\cos KX] = \frac{P}{P^2 + K^2}$$

(4) the following ex ample illustrates
the relationships between
the exponential function
the hyperbolic functions
and
their Laplace transforms

Chapter 14 The Inverse Laplace Transform

Find the inverse Laplace transform of
$$F(p) = \frac{1}{p^2 - 1}$$
using hyperbolic function and exponential function

Hyperbolic functions

$$F(p) = \frac{1}{p^2 - 1}$$

$$L^{-1}\left[\frac{K}{p^2 - K^2}\right] = \sinh Kx$$

$$L^{-1}\left[\frac{1}{p^2 - 1}\right]$$

$$= L^{-1}\left[\frac{1}{p^2 - 1^2}\right]$$

$$= \sinh x \qquad\qquad K = 1$$

or $f(x) = \sinh x$

Chapt 14 The Inverse Laplace Transform

exponential functions

$$F(p) = \frac{1}{p^2 - 1}$$

$$L^{-1}\left[\frac{1}{p-K}\right] = e^{Kx}$$

$$= \frac{\dfrac{1}{p^2-1}}{\dfrac{1}{(p-1)(p+1)}}$$

$$\frac{1}{(p-1)(p+1)} = \frac{A}{p-1} + \frac{B}{p+1}$$

$$= \frac{A}{p-1}\frac{p+1}{p+1} + \frac{B}{p+1}\frac{p-1}{p-1}$$

$$= \frac{A(p+1) + B(p-1)}{(p-1)(p+1)}$$

$$1 = A(p+1) + B(p-1)$$
$$A(p+1) + B(p-1) = 1$$
$$Ap + A + Bp - B = 1$$
$$Ap + Bp + A - B = 1$$
$$(A + B)p + (A - B) = 1$$
$$A + B = 0 \qquad A - B = 1$$

(198)

Chapt 14 The Inverse Laplace Transform

$A + B = 0$

$A - B = 1$

$$A = \frac{\begin{vmatrix} 0 & 1 \\ 1 & -1 \end{vmatrix}}{\begin{vmatrix} 1 & 1 \\ 1 & -1 \end{vmatrix}} = \frac{0 - 1}{-1 - 1} = \frac{-1}{-2} = \frac{1}{2}$$

$$B = \frac{\begin{vmatrix} 1 & 0 \\ 1 & 1 \end{vmatrix}}{\begin{vmatrix} 1 & 1 \\ 1 & -1 \end{vmatrix}} = \frac{1 - 0}{-1 - 1} = \frac{1}{-2} = -\frac{1}{2}$$

$A = \frac{1}{2}$, $B = -\frac{1}{2}$

$$\frac{\frac{1}{2}}{p-1} + \frac{-\frac{1}{2}}{p+1}$$

$$\frac{\frac{1}{2}}{p-1} - \frac{\frac{1}{2}}{p+1}$$

Chapt 14 The Inverse Laplace Transform

$$L^{-1}\left[\frac{1}{p^2 - 1}\right]$$

$$= L^{-1}\left[\frac{1}{(p-1)(p+1)}\right]$$

$$= L^{-1}\left[\frac{\frac{1}{2}}{p-1} - \frac{\frac{1}{2}}{p+1}\right]$$

$$= L^{-1}\left[\frac{\frac{1}{2}}{p-1}\right] - L^{-1}\left[\frac{\frac{1}{2}}{p+1}\right]$$

$$= \frac{1}{2} L^{-1}\left[\frac{1}{p-1}\right] - \frac{1}{2} L^{-1}\left[\frac{1}{p+1}\right]$$

$$= \frac{1}{2} e^{x} - \frac{1}{2} e^{-x}$$

(200)

Chapter 14 The Inverse Laplace Transform

note

the two answers are equivalent

$$f(x) = \sinh x$$

and

$$f(x) = \frac{1}{2}e^x - \frac{1}{2}e^{-x}$$

$$= \frac{1}{2}(e^x - e^{-x})$$

$$= \frac{e^x - e^{-x}}{2}$$

$$= \sinh x$$

(201)

Chapt 14 The Inverse Laplace Transform
Preference For Using The Exponential Function
And Partial Fractions.

(1) when encountering a Laplace transform of the form

$$F_{(p)} = \frac{K}{p^2 - k^2} \qquad or$$

$$F_{(p)} = \frac{p}{p^2 - k^2}$$

there is a preference to find the corresponding function $f(x)$ by

(1) factoring the denominator   and

(2) using partial fractions   and

(3) expressing $f(x)$ in terms of exponential functions

I

(202)

Chapt 14 The Inverse Laplce Transform

① given $F(p) = \dfrac{1}{(p-3)^2}$

find $f(x)$

$e^{ax} \quad x^n$

$$F(p) = \frac{1}{(p-3)^2}$$

$$L[f(x)] = \frac{1}{(p-3)^2}$$

$$L^{-1}[L[f(x)]] = L^{-1}\left[\frac{1}{(p-3)^2}\right]$$

$$f(x) = L^{-1}\left[\frac{1}{(p-3)^2}\right]$$

$$f(x) = L^{-1}\left[\frac{1!}{(p-3)^{1+1}}\right]$$

$$f(x) = x^1 \, e^{3x}$$

$$f(x) = x \, e^{3x}$$

Chapter 14 The Inverse Laplace Transform

1. given $F(p) = \dfrac{2}{(p-5)^3}$

find $f(x)$

$e^{ax} \; x^n$

$$F(p) = \frac{2}{(p-5)^3}$$

$$L[f(x)] = \frac{2}{(p-5)^3}$$

$$L^{-1}[L[f(x)]] = L^{-1}\left[\frac{2}{(p-5)^3}\right]$$

$$f(x) = L^{-1}\left[\frac{2!}{(p-5)^{2+1}}\right]$$

$$f(x) = x^2 e^{5x}$$

Chapt 14 The Inverse Laplace Transform

(204)

(3) given $F(p) = \dfrac{1}{(p+2)^2}$

find $f(x)$

x^{ax} x^n

$F(p) = \dfrac{1}{(p+2)^2}$

$L[f(x)] = \dfrac{1}{(p+2)^2}$

$L^{-1}[L[f(x)]] = L^{-1}\left[\dfrac{1}{(p+2)^2}\right]$

$f(x) = L^{-1}\left[\dfrac{1}{(p+2)^2}\right]$

$f(x) = L^{-1}\left[\dfrac{1!}{(p+2)^{1+1}}\right]$

$f(x) = x^1 e^{-2x}$

$f(x) = x e^{-2x}$

(205)

Chapt 14 The Inverse Laplace Transform

(4) given $F_{(p)} = \dfrac{2}{(p+4)^3}$

find $f(x)$

$\mathcal{L}^{ax} \, x^n$

$F(p) = \dfrac{2}{(p+4)^3}$

$L[f(x)] = \dfrac{2}{(p+4)^3}$

$L^{-1}[L[f(x)]] = L^{-1}\left[\dfrac{2}{(p+4)^3}\right]$

$f(x) = L^{-1}\left[\dfrac{2}{(p+4)^3}\right]$

$f(x) = L^{-1}\left[\dfrac{2!}{(p+4)^{2+1}}\right]$

$f(x) = x^2 \, e^{-4x}$

$\overline{V}$

PROB

(206)

Chapt 14 The Inverse Laplace Transform

(5) given $\quad F(p) = \dfrac{1}{(p+1)^4}$

find $f(x)$

$e^{ax} \quad x^n$

$F(p) = \dfrac{1}{(p+1)^4}$

$L[f(x)] = \dfrac{1}{(p+1)^4}$

$L^{-1}[L[f(x)]] = L^{-1}\left[\dfrac{1}{(p+1)^4}\right]$

$f(x) = L^{-1}\left[\dfrac{1}{(p+1)^4}\right]$

$f(x) = L^{-1}\left[\dfrac{1}{(p+1)^{3+1}}\right]$

$f(x) = L^{-1}\left[\dfrac{\frac{3!}{3!}}{(p+1)^{3+1}}\right]$

$f(x) = \dfrac{1}{3!} L^{-1}\left[\dfrac{3!}{(p+1)^{3+1}}\right]$

$f(x) = \dfrac{1}{6} x^3 e^{-x}$

Chapt 14 The Inverse Laplace Transform $\qquad$ (207)

⑥ given $F(p) = \dfrac{7}{(p+3)^3}$

find $f(x)$

$\underline{e^{ax} \, x^n}$

$$F(p) = \dfrac{7}{(p+3)^3}$$

$$L[f(x)] = \dfrac{7}{(p+3)^3}$$

$$L^{-1}\left[L[f(x)]\right] = L^{-1}\left[\dfrac{7}{(p+3)^3}\right]$$

$$f(x) = L^{-1}\left[\dfrac{7}{(p+3)^3}\right]$$

$$f(x) = 7\, L^{-1}\left[\dfrac{1}{(p+3)^3}\right]$$

$$f(x) = 7\, L^{-1}\left[\dfrac{1}{(p+3)^{2+1}}\right]$$

$$f(x) = 7\, L^{-1}\left[\dfrac{\frac{2!}{2!}}{(p+3)^{2+1}}\right]$$

$$f(x) = \dfrac{7}{2!}\, L^{-1}\left[\dfrac{2!}{(p+3)^{2+1}}\right]$$

$$f(x) = \dfrac{7}{2}\, x^2\, e^{-3x}$$

I

Chapt 14 The Inverse Laplace Transform $\textcircled{208}$

① evaluate

$$L^{-1}\left[\frac{1}{(p-3)^2+1}\right]$$

$\underline{e^{ax}\,\sin bx}$

$$L^{-1}\left[\frac{1}{(p-3)^2+1}\right]$$

$$= L^{-1}\left[\frac{1}{(p-3)^2+1^2}\right]$$

$$= \sin x \; e^{3x}$$

$$= e^{3x}\,\sin x$$

② evaluate

$$L^{-1}\left[\frac{2}{(p-5)^2+4}\right]$$

$$L^{-1}\left[\frac{2}{(p-5)^2+4}\right]$$

$$= L^{-1}\left[\frac{2}{(p-5)^2+2^2}\right]$$

$$= \sin 2x \; e^{5x}$$

$$= e^{5x}\,\sin 2x$$

Chpt 14 The Inverse Laplace Transform (209)

③ evaluate

$$L^{-1}\left[\frac{3}{(p+2)^2+9}\right]$$

$\dfrac{e^{ax}}{\sin bx}$

$$L^{-1}\left[\frac{3}{(p+2)^2+9}\right]$$

$$=L^{-1}\left[\frac{3}{(p+2)^2+3^2}\right]$$

$$= \sin 3x \quad e^{-2x}$$

$$= e^{-2x}\sin 3x$$

④ evaluate

$$L^{-1}\left[\frac{5}{(p+1)^2+4}\right]$$

$$L^{-1}\left[\frac{5}{(p+1)^2+4}\right]$$

$$=L^{-1}\left[\frac{5}{(p+1)^2+2^2}\right]$$

$$=L^{-1}\left[\frac{5\cdot\frac{2}{2}}{(p+1)^2+2^2}\right]$$

$$=\frac{5}{2}L^{-1}\left[\frac{2}{(p+1)^2+2^2}\right]$$

$$=\frac{5}{2}\sin 2x\, e^{-x}=\frac{5}{2}e^{-x}\sin 2x$$

Chapt 14 The Inverse Laplace Transform (2-10)

① evaluate

$$L^{-1}\left[\frac{p-2}{(p-2)^2 + 1}\right]$$

$\frac{e^{ax}}{\cos bx}$

$$L^{-1}\left[\frac{p-2}{(p-2)^2 + 1}\right]$$

$$= L^{-1}\left[\frac{p-2}{(p-2)^2 + 1^2}\right]$$

$$= \cos x \; e^{2x}$$

$$= e^{2x}\cos x$$

② evaluate

$$L^{-1}\left[\frac{p-3}{(p-3)^2 + 4}\right]$$

$$L^{-1}\left[\frac{p-3}{(p-3)^2 + 4}\right]$$

$$= L^{-1}\left[\frac{p-3}{(p-3)^2 + 2^2}\right]$$

$$= \cos 2x \; e^{3x}$$

$$= e^{3x}\cos 2x$$

Chapt 14 The Inverse Laplace Transfor- $\Sigma 11$

③ evaluate

$$L^{-1}\left[\frac{p+4}{(p+4)^2 + 25}\right]$$

$\varepsilon^{ax} \cos bx$

$$L^{-1}\left[\frac{p+4}{(p+4)^2 + 25}\right]$$

$$= L^{-1}\left[\frac{p+4}{(p+4)^2 + 5^2}\right]$$

$$= \cos 5x \quad \varepsilon^{-4x}$$

$$= \varepsilon^{-4x} \cos 5x$$

④ evaluate

$$L^{-1}\left[\frac{5p+5}{(p+1)^2 + 16}\right]$$

$$L^{-1}\left[\frac{5p+5}{(p+1)^2 + 16}\right]$$

$$= L^{-1}\left[\frac{5(p+1)}{(p+1)^2 + 4^2}\right]$$

$$= 5 L^{-1}\left[\frac{p+1}{(p+1)^2 + 4^2}\right]$$

$$= 5 \cos 4x \quad \varepsilon^{-x}$$

$$= 5 \varepsilon^{-x} \cos 4x$$

(212)

Chapt 15 The Inverse Laplace Transform Operator to Linear Linear Operators

① an operator T is linear if it has the following two properties

① $T[a\,f(x)] = a\,T[f(x)]$ and

② $T[f(x) + g(x)] = T[f(x)] + T[g(x)]$

② from these two properties follow a third property

$$T[a\,f(x) + b\,g(x)] = T[a\,f(x)] + T[b\,g(x)]$$
$$= a\,T[f(x)] + b\,T[g(x)]$$

note

① $y = 2x$ is a function — a relationship between two variables x and y therefore we can write

$f(x) = 2x$

② in general we can write

$f(x) = Kx$ where K is a constant

③ however with regard to the above two properties we like to be as explicit as possible

chapt 15 The Inverse Laplace Transform Operator is Linear

The Inverse Laplace Transform Operator
is Linear

① the inverse Laplace transform operator L^{-1}
is a linear operator

② therefore the operator has the following properties
$$L^{-1}[a f(x)] = a L^{-1}[f(x)]$$
$$L^{-1}[f(x) + g(x)] = L^{-1}[f(x)] + L^{-1}[g(x)]$$

③ from these two properties follow a third property
$$L^{-1}[a f(x) + b g(x)] = L^{-1}[a f(x)] + L^{-1}[b g(x)]$$
$$= a L^{-1}[f(x)] + b L^{-1}[g(x)]$$

④ the linear properties of
the inverse Laplace transform operator L^{-1}
allow us to find the inverse transforms of
more complex functions involving
a sum of functions
a difference of functions
and
a constant times a function

⑤ the following examples illustrate the linear properties
of the inverse Laplace transform operator L^{-1}

(214)

Chapt 15 The Inverse Laplace Transform Operator Is Linear

ex find the inverse Laplace Transform of

$$F(p) = \frac{1}{p^2} + \frac{2}{p^3}$$

$$F(p) = \frac{1}{p^2} + \frac{2}{p^3}$$

$$L[\theta(x)] = \frac{1}{p^2} + \frac{2}{p^3}$$

$$L^{-1}[L[\theta(x)]] = L^{-1}\left[\frac{1}{p^2} + \frac{2}{p^3}\right]$$

$$\theta(x) = L^{-1}\left[\frac{1}{p^2}\right] + L^{-1}\left[\frac{2}{p^3}\right]$$

$$\theta(x) = L^{-1}\left[\frac{1!}{p^{1+1}}\right] + L^{-1}\left[\frac{2!}{p^{2+1}}\right]$$

$$\theta(x) = x^1 + x^2$$
$$\theta(x) = x + x^2$$
$$\theta(x) = x^2 + x$$

(215)

Chapt 15 The Inverse Laplace Transform Operates of Linear

Ex) find the inverse Laplace Transform of

$$F(p) = \frac{3}{p^2} + \frac{14}{p^3}$$

$$F(p) = \frac{3}{p^2} + \frac{14}{p^3}$$

$$L[f(x)] = \frac{3}{p^2} + \frac{14}{p^3}$$

$$L^{-1}[L[f(x)]] = L^{-1}\left[\frac{3}{p^2} + \frac{14}{p^3}\right]$$

$$f(x) = L^{-1}\left[\frac{3}{p^2}\right] + L^{-1}\left[\frac{14}{p^3}\right]$$

$$f(x) = L^{-1}\left[\frac{3}{p^{1+1}}\right] + L^{-1}\left[\frac{14}{p^{2+1}}\right]$$

$$f(x) = L^{-1}\left[\frac{3 \cdot 1}{p^{1+1}}\right] + L^{-1}\left[\frac{7 \cdot 2}{p^{2+1}}\right]$$

$$f(x) = 3 L^{-1}\left[\frac{1}{p^{1+1}}\right] + 7 L^{-1}\left[\frac{2}{p^{2+1}}\right]$$

$$f(x) = 3 L^{-1}\left[\frac{1!}{p^{1+1}}\right] + 7 L^{-1}\left[\frac{2!}{p^{2+1}}\right]$$

$$f(x) = 3 \cdot x^1 + 7 \cdot x^2$$

$$f(x) = 3x + 7x^2$$

$$f(x) = 7x^2 + 3x$$

Chapt 15 The Inverse Laplace Transform Operation is Linear

ex   find the inverse Laplace transform of

$$F(p) = \frac{2}{p} - \frac{4}{p^3} + \frac{18}{p^4}$$

$$F(p) = \frac{2}{p} - \frac{4}{p^3} + \frac{18}{p^4}$$

$$L[f(x)] = \frac{2}{p} - \frac{4}{p^3} + \frac{18}{p^4}$$

$$L^{-1}[L[f(x)]] = L^{-1}\left[\frac{2}{p} - \frac{4}{p^3} + \frac{18}{p^4}\right]$$

$$f(x) = L^{-1}\left[\frac{2}{p}\right] - L^{-1}\left[\frac{4}{p^3}\right] + L^{-1}\left[\frac{18}{p^4}\right]$$

$$f(x) = L^{-1}\left[\frac{2}{p}\right] - L^{-1}\left[\frac{4}{p^{2+1}}\right] + L^{-1}\left[\frac{18}{p^{3+1}}\right]$$

$$f(x) = L^{-1}\left[\frac{2}{p}\right] - L^{-1}\left[\frac{2 \cdot 2}{p^{2+1}}\right] + L^{-1}\left[\frac{3 \cdot 6}{p^{3+1}}\right]$$

$$f(x) = 2 L^{-1}\left[\frac{1}{p}\right] - 2 L^{-1}\left[\frac{2}{p^{2+1}}\right] + 3 L^{-1}\left[\frac{6}{p^{3+1}}\right]$$

$$f(x) = 2 L^{-1}\left[\frac{1}{p}\right] - 2 L^{-1}\left[\frac{2!}{p^{2+1}}\right] + 3 L^{-1}\left[\frac{3!}{p^{3+1}}\right]$$

$$f(x) = 2 \cdot 1 - 2 \cdot x^2 + 3 \cdot x^3$$
$$f(x) = 2 - 2x^2 + 3x^3$$
$$f(x) = 3x^3 - 2x^2 + 2$$

Chapter 15 The Inverse Laplace Transform Operator to Lunes

ex find the inverse Laplace transform of

$$F(p) = \frac{3}{p} - \frac{6}{p^4} + \frac{2}{p^2+4} - \frac{p}{p^2+9}$$

$$F(p) = \frac{3}{p} - \frac{6}{p^4} + \frac{2}{p^2+4} - \frac{p}{p^2+9}$$

$$L[f(x)] = \frac{3}{p} - \frac{6}{p^4} + \frac{2}{p^2+4} - \frac{p}{p^2+9}$$

$$L^{-1}[L[f(x)]] = L^{-1}\left[\frac{3}{p} - \frac{6}{p^4} + \frac{2}{p^2+4} - \frac{p}{p^2+9}\right]$$

$$f(x) = L^{-1}\left[\frac{3}{p}\right] - L^{-1}\left[\frac{6}{p^4}\right] + L^{-1}\left[\frac{2}{p^2+4}\right] - L^{-1}\left[\frac{p}{p^2+9}\right]$$

$$f(x) = L^{-1}\left[\frac{3}{p}\right] - L^{-1}\left[\frac{3!}{p^{3+1}}\right] + L^{-1}\left[\frac{2}{p^2+2^2}\right] - L^{-1}\left[\frac{p}{p^2+3^2}\right]$$

$$f(x) = 3 - x^3 + \sin 2x - \cos 3x$$

Chapter 15 The Inverse Laplace Transform Operator to Linear

Ex. find the inverse Laplace Transform of

$$F(p) = \frac{5}{p^5} + \frac{7}{p-3}$$

$$F(p) = \frac{5}{p^5} + \frac{7}{p-3}$$

$$L[f(x)] = \frac{5}{p^5} + \frac{7}{p-3}$$

$$L^{-1}[L[f(x)]] = L^{-1}\left[\frac{5}{p^5} + \frac{7}{p-3}\right]$$

$$f(x) = L^{-1}\left[\frac{5}{p^5}\right] + L^{-1}\left[\frac{7}{p-3}\right]$$

$$f(x) = L^{-1}\left[\frac{5}{p^{4+1}}\right] + 7 L^{-1}\left[\frac{1}{p-3}\right]$$

$$f(x) = L^{-1}\left[\frac{5 \cdot \frac{4!}{4!}}{p^{4+1}}\right] + 7 e^{3x}$$

$$f(x) = \frac{5}{4!} L^{-1}\left[\frac{4!}{p^{4+1}}\right] + 7 e^{3x}$$

$$f(x) = \frac{5}{4 \cdot 3 \cdot 2 \cdot 1} x^4 + 7 e^{3x}$$

$$f(x) = \frac{5}{24} x^4 + 7 e^{3x}$$

(219)

Chapt 15 The Inverse Laplace Transform Operation / Linear

Q) evaluate

$$L^{-1}\left[\frac{5}{p^2-1}\right]$$

expand) $$L^{-1}\left[\frac{5}{p^2-1}\right]$$

$$= L^{-1}\left[\frac{5}{(p-1)(p+1)}\right]$$

$$\frac{5}{(p-1)(p+1)} = \frac{A}{p-1} + \frac{B}{p+1}$$

$$= \frac{A}{p-1}\frac{p+1}{p+1} + \frac{B}{p+1}\frac{p-1}{p-1}$$

$$= \frac{A(p+1) + B(p-1)}{(p-1)(p+1)}$$

$$5 = A(p+1) + B(p-1)$$
$$A(p+1) + B(p-1) = 5$$
$$Ap + A + Bp - B = 5$$
$$Ap + Bp + A - B = 5$$
$$(A+B)p + (A-B) = 5$$
$$A+B = 0 \qquad A-B = 5$$

$$A + B = 0$$
$$A - B = 5$$

$$\boxed{220}$$

Chapt 15 The Inverse Laplace Operator to Linear

(1 cont)

$$A + B = 0$$
$$A - B = 5$$

exponential

$$A = \frac{\begin{vmatrix} 0 & 1 \\ 5 & -1 \end{vmatrix}}{\begin{vmatrix} 1 & 1 \\ 1 & -1 \end{vmatrix}} = \frac{0-5}{-1-1} = \frac{-5}{-2} = \frac{5}{2}$$

$$B = \frac{\begin{vmatrix} 1 & 0 \\ 1 & 5 \end{vmatrix}}{\begin{vmatrix} 1 & 1 \\ 1 & -1 \end{vmatrix}} = \frac{5-0}{-1-1} = \frac{5}{-2} = -\frac{5}{2}$$

$$A = \frac{5}{2}, \quad B = -\frac{5}{2}$$

$$\frac{A}{p-1} + \frac{B}{p+1}$$

$$= \frac{\frac{5}{2}}{p-1} + \frac{-\frac{5}{2}}{p+1}$$

$$= \frac{\frac{5}{2}}{p-1} - \frac{\frac{5}{2}}{p+1}$$

IC

2 21

(cont)

$$L^{-1}\left[\frac{5}{p^2-1}\right]$$

$$= L^{-1}\left[\frac{5}{(p-1)(p+1)}\right]$$

exponential $= L^{-1}\left[\frac{\frac{5}{2}}{p-1} - \frac{\frac{5}{2}}{p+1}\right]$

$$= L^{-1}\left[\frac{\frac{5}{2}}{p-1}\right] - L^{-1}\left[\frac{\frac{5}{2}}{p+1}\right]$$

$$= \frac{5}{2}L^{-1}\left[\frac{1}{p-1}\right] - \frac{5}{2}L^{-1}\left[\frac{1}{p+1}\right]$$

$$= \frac{5}{2}e^x - \frac{5}{2}e^{-x}$$

227

(1cont) also 0   using hyperbolic functions

$$L^{-1}\left[\frac{5}{p^2 - 1}\right]$$

Exponential
$$= 5 \ L^{-1}\left[\frac{1}{p^2 - 1}\right]$$

$$= 5 \ L^{-1}\left[\frac{1}{p^2 - 1^2}\right]$$

$$= 5 \ \sinh x$$

$$= 5 \left(\frac{e^x - e^{-x}}{2}\right)$$

$$= \frac{5}{2} \left(e^x - e^{-x}\right)$$

$$= \frac{5}{2} e^x - \frac{5}{2} e^{-x}$$

II A   Prob

Chapt 15 The Inverse Laplace Transform Operator to Linear

② evaluate

$$L^{-1}\left[\frac{7}{p^2 - 4}\right]$$

exponential

$$L^{-1}\left[\frac{7}{p^2 - 4}\right]$$

$$= L^{-1}\left[\frac{7}{(p-2)(p+2)}\right]$$

$$\frac{7}{(p-2)(p+2)} = \frac{A}{p-2} + \frac{B}{p+2}$$

$$= \frac{A}{p-2}\,\frac{p+2}{p+2} + \frac{B}{p+2}\,\frac{p-2}{p-2}$$

$$= \frac{A(p+2) + B(p-2)}{(p-2)(p+2)}$$

$$7 = A(p+2) + B(p-2)$$
$$A(p+2) + B(p-2) = 7$$
$$Ap + 2A + Bp - 2B = 7$$
$$Ap + Bp + 2A - 2B = 7$$
$$(A+B)p + (2A - 2B) = 7$$
$$A + B = 0 \qquad 2A - 2B = 7$$

$$A + B = 0$$
$$2A - 2B = 7$$

$$\boxed{224}$$

Chapt 15 The Inverse Laplace Transform Operator Volume

(2 cont)

$$A + B = 0$$
$$2A - 2B = 7$$

exponential

$$A = \frac{\begin{vmatrix} 0 & 1 \\ 7 & -2 \end{vmatrix}}{\begin{vmatrix} 1 & 1 \\ 2 & -2 \end{vmatrix}} = \frac{0-7}{-2-2} = \frac{-7}{-4} = \frac{7}{4}$$

$$B = \frac{\begin{vmatrix} 1 & 0 \\ 2 & 7 \end{vmatrix}}{\begin{vmatrix} 1 & 1 \\ 2 & -2 \end{vmatrix}} = \frac{7-0}{-2-2} = \frac{7}{-4} = -\frac{7}{4}$$

$$\frac{A}{p-2} + \frac{B}{p+2}$$

$$= \frac{\frac{7}{4}}{p-2} + \frac{-\frac{7}{4}}{p+2}$$

$$= \frac{\frac{7}{4}}{p-2} - \frac{\frac{7}{4}}{p+2}$$

Chapt 15 The Inverse Laplace Transform ~ Operator to Laws

(2 Cont) $L^{-1}\left[\dfrac{7}{p^2 - 4}\right]$

$= L^{-1}\left[\dfrac{7}{(p-2)(p+2)}\right]$

expand $= L^{-1}\left[\dfrac{\frac{7}{4}}{p-2} - \dfrac{\frac{7}{4}}{p+2}\right]$

$= L^{-1}\left[\dfrac{\frac{7}{4}}{p-2}\right] - L^{-1}\left[\dfrac{\frac{7}{4}}{p+2}\right]$

$= \dfrac{7}{4} \, L^{-1}\left[\dfrac{1}{p-2}\right] - \dfrac{7}{4} \, L^{-1}\left[\dfrac{1}{p+2}\right]$

$= \dfrac{7}{4} \, e^{2x} - \dfrac{7}{4} \, e^{-2x}$

Chapt 15 The Inverse Laplace Transform Operation to Lower algo   using hyperbolic function

(2 cont)

$$L^{-1}\left[\frac{7}{p^2 - 4}\right]$$

$$= L^{-1}\left[\frac{7}{p^2 - 2^2}\right]$$

exponential

$$= L^{-1}\left[\frac{7 \cdot \frac{2}{2}}{p^2 - 2^2}\right]$$

$$= \frac{7}{2} L^{-1}\left[\frac{2}{p^2 - 2^2}\right]$$

$$= \frac{7}{2} \sinh 2x$$

$$= \frac{7}{2}\left(\frac{e^{2x} - e^{-2x}}{2}\right)$$

$$= \frac{7}{4}\left(e^{2x} - e^{-2x}\right)$$

$$= \frac{7}{4} e^{2x} - \frac{7}{4} e^{-2x}$$

227

③ Chapt 15 The Inverse Laplace Transform Operator to Linear

③ Evaluate

$$L^{-1}\left[\frac{3}{p^2 - 3p + 2}\right]$$

exponential

$$L^{-1}\left[\frac{3}{p^2 - 3p + 2}\right]$$

$$= L^{-1}\left[\frac{3}{(p-1)(p-2)}\right]$$

$$\frac{3}{(p-1)(p-2)} = \frac{A}{p-1} + \frac{B}{p-2}$$

$$= \frac{A}{p-1}\frac{p-2}{p-2} + \frac{B}{p-2}\frac{p-1}{p-1}$$

$$= \frac{A(p-2) + B(p-1)}{(p-1)(p-2)}$$

$$3 = A(p-2) + B(p-1)$$

$$A(p-2) + B(p-1) = 3$$

$$Ap - 2A + Bp - B = 3$$

$$Ap + Bp - 2A - B = 3$$

$$(A + B)p + (-2A - B) = 3$$

$$A + B = 0 \qquad -2A - B = 3$$

$$2A + B = -3$$

III B

Prob

$\Sigma 228$

Chapt 15 The Inverse Laplace Transform - Operator for Linear

(3 cont)

$A + B = 0$

$2A + B = -3$

exponential

$$A = \frac{\begin{vmatrix} 0 & 1 \\ -3 & 1 \end{vmatrix}}{\begin{vmatrix} 1 & 1 \\ 2 & 1 \end{vmatrix}} = \frac{0 + 3}{1 - 2} = \frac{3}{-1} = -3$$

$$B = \frac{\begin{vmatrix} 1 & 0 \\ 2 & -3 \end{vmatrix}}{\begin{vmatrix} 1 & 1 \\ 2 & 1 \end{vmatrix}} = \frac{-3 - 0}{1 - 2} = \frac{-3}{-1} = 3$$

$A = -3, \quad B = 3$

$$\frac{A}{p-1} + \frac{B}{p-2}$$

$$= \frac{-3}{p-1} + \frac{3}{p-2}$$

$$= \frac{3}{p-2} - \frac{3}{p-1}$$

(229)

Ch wt 15 The Inverse Laplace Transform Operator to know

(3 cont) $L^{-1}\left[\dfrac{3}{p^2 - 3p + 2}\right]$

$= L^{-1}\left[\dfrac{3}{(p-1)(p-2)}\right]$

x expand $= L^{-1}\left[\dfrac{3}{p-2} - \dfrac{3}{p-1}\right]$

$= L^{-1}\left[\dfrac{3}{p-2}\right] - L^{-1}\left[\dfrac{3}{p-1}\right]$

$= 3\, L^{-1}\left[\dfrac{1}{p-2}\right] - 3\, L^{-1}\left[\dfrac{1}{p-1}\right]$

$= 3\, e^{2x} - 3\, e^{x}$

note

the inverse Laplace transform $f(x)$
cannot be expressed in terms of
sinh x or cosh x
$f(x) = 3\, e^{2x} - 3\, e^{x}$

(230)

Chapt 15 The Inverse Laplace Transform Operator to Linear

④ evaluate

$$L^{-1}\left[\frac{p+5}{p^2 - 2p - 3}\right]$$

exponential $\quad L^{-1}\left[\dfrac{p+5}{p^2 - 2p - 3}\right]$

$$= L^{-1}\left[\frac{p+5}{(p-3)(p+1)}\right]$$

$$= L^{-1}\left[\frac{p+5}{(p+1)(p-3)}\right]$$

$$\frac{p+5}{(p+1)(p-3)} = \frac{A}{p+1} + \frac{B}{p-3}$$

$$= \frac{A}{p+1}\frac{p-3}{p-3} + \frac{B}{p-3}\frac{p+1}{p+1}$$

$$= \frac{A(p-3) + B(p+1)}{(p+1)(p-3)}$$

$$p + 5 = A(p-3) + B(p+1)$$

$$A(p-3) + B(p+1) = p+5$$

$$Ap - 3A + Bp + B = p+5$$

$$Ap + Bp - 3A + B = p+5$$

$$(A+B)p + (-3A + B) = p+5$$

$$A + B = 1 \qquad -3A + B = 5$$

$$3A - B = -5$$

$\boxed{231}$

Chapter 15 The Inverse Laplace Transform Operator to Linear

$A + B = 1$

$3A - B = -5$

exponential $A = \dfrac{\begin{vmatrix} 1 & 1 \\ -5 & -1 \end{vmatrix}}{\begin{vmatrix} 1 & 1 \\ 3 & -1 \end{vmatrix}} = \dfrac{-1 + 5}{-1 - 3} = \dfrac{4}{-4} = -1$

$B = \dfrac{\begin{vmatrix} 1 & 1 \\ 3 & -5 \end{vmatrix}}{\begin{vmatrix} 1 & 1 \\ 3 & -1 \end{vmatrix}} = \dfrac{-5 - 3}{-1 - 3} = \dfrac{-8}{-4} = 2$

$A = -1, \quad B = 2$

$\dfrac{A}{p+1} + \dfrac{B}{p-3}$

$= \dfrac{-1}{p+1} + \dfrac{2}{p-3}$

$= \dfrac{2}{p-3} - \dfrac{1}{p+1}$

Chpt 15 The Inverse Laplace Transform - Operator $\forall$. Line

(4 cont)
$$L^{-1}\left[\frac{p+5}{p^2-2p-3}\right]$$

$$= L^{-1}\left[\frac{p+5}{(p+1)(p-3)}\right]$$

exponential $= L^{-1}\left[\frac{2}{p-3} - \frac{1}{p+1}\right]$

$$= L^{-1}\left[\frac{2}{p-3}\right] - L^{-1}\left[\frac{1}{p+1}\right]$$

$$= 2\,L^{-1}\left[\frac{1}{p-3}\right] - L^{-1}\left[\frac{1}{p+1}\right]$$

$$= 2\,e^{3x} - e^{-x}$$

(233)

⑤ evaluate

Chrt 15 The Inverse Laplace Transform Operation γ_1 Laws

⑤ evaluate

$$L^{-1}\left[\frac{3p+7}{p^2-9}\right]$$

exponential

$$L^{-1}\left[\frac{3p+7}{p^2-9}\right]$$

$$= L^{-1}\left[\frac{3p+7}{(p-3)(p+3)}\right]$$

$$\frac{3p+7}{(p-3)(p+3)} = \frac{A}{p-3} + \frac{B}{p+3}$$

$$= \frac{A}{p-3}\cdot\frac{p+3}{p+3} + \frac{B}{p+3}\cdot\frac{p-3}{p-3}$$

$$= \frac{A(p+3)+B(p-3)}{(p-3)(p+3)}$$

$$3p+7 = A(p+3) + B(p-3)$$

$$A(p+3) + B(p-3) = 3p+7$$

$$Ap + 3A + Bp - 3B = 3p+7$$

$$Ap + Bp + 3A - 3B = 3p+7$$

$$(A+B)p + (3A-3B) = 3p+7$$

$$A+B = 3 \qquad 3A-3B = 7$$

$$A+B = 3$$
$$3A - 3B = 7$$

$\overline{\text{V}}$ B

(234)

Chapter 15 The Inverse Laplace Transform Operator Volume

(5 cont)
$$A + B = 3$$
$$3A - 3B = 7$$

exponential

$$A = \frac{\begin{vmatrix} 3 & 1 \\ 7 & -3 \end{vmatrix}}{\begin{vmatrix} 1 & 1 \\ 3 & -3 \end{vmatrix}} = \frac{-9-7}{-3-3} = \frac{-16}{-6} = \frac{8}{3}$$

$$B = \frac{\begin{vmatrix} 1 & 3 \\ 3 & 7 \end{vmatrix}}{\begin{vmatrix} 1 & 1 \\ 3 & -3 \end{vmatrix}} = \frac{7-9}{-3-3} = \frac{-2}{-6} = \frac{1}{3}$$

$$A = \frac{8}{3}, \quad B = \frac{1}{3}$$

$$= \frac{A}{p-3} + \frac{B}{p+3}$$
$$= \frac{\frac{8}{3}}{p-3} + \frac{\frac{1}{3}}{p+3}$$

Chapt 15 The Inverse Laplace Transform — Operation of Linear

(5 cont)

$$L^{-1}\left[\frac{3p+7}{p^2-9}\right]$$

$$= L^{-1}\left[\frac{3p+7}{(p-3)(p+3)}\right]$$

expand f() $= L^{-1}\left[\frac{\frac{8}{3}}{p-3} + \frac{\frac{1}{3}}{p+3}\right]$

$$= L^{-1}\left[\frac{\frac{8}{3}}{p-3}\right] + L^{-1}\left[\frac{\frac{1}{3}}{p+3}\right]$$

$$= \frac{8}{3}L^{-1}\left[\frac{1}{p-3}\right] + \frac{1}{3}L^{-1}\left[\frac{1}{p+3}\right]$$

$$= \frac{8}{3}e^{3x} + \frac{1}{3}e^{-3x}$$

(236)

Chapt 15 The Inverse Laplace Transform Applications to General

(5 cont) also 0   using hyperbola functions

$$L^{-1}\left[\frac{3p+7}{p^2-9}\right]$$

exponential

$$= L^{-1}\left[\frac{3p}{p^2-9}+\frac{7}{p^2-9}\right]$$

$$= L^{-1}\left[\frac{3p}{p^2-9}\right]+L^{-1}\left[\frac{7}{p^2-9}\right]$$

$$= 3\,L^{-1}\left[\frac{p}{p^2-9}\right]+7\,L^{-1}\left[\frac{1}{p^2-9}\right]$$

$$= 3\,L^{-1}\left[\frac{p}{p^2-3^2}\right]+7\,L^{-1}\left[\frac{1}{p^2-3^2}\right]$$

$$= 3\,L^{-1}\left[\frac{p}{p^2-3^2}\right]+7\,L^{-1}\left[\frac{\frac{3}{3}}{p^2-3^2}\right]$$

$$= 3\,L^{-1}\left[\frac{p}{p^2-3^2}\right]+\frac{7}{3}\,L^{-1}\left[\frac{3}{p^2-3^2}\right]$$

$$= 3\cosh 3x+\frac{7}{3}\sinh 3x$$

$$= 3\left(\frac{e^{3x}+e^{-3x}}{2}\right)+\frac{7}{3}\left(\frac{e^{3x}-e^{-3x}}{2}\right)$$

$$= \frac{3}{2}\left(e^{3x}+e^{-3x}\right)+\frac{7}{6}\left(e^{3x}-e^{-3x}\right)$$

$$= \frac{9}{6}\left(e^{3x}+e^{-3x}\right)+\frac{7}{6}\left(e^{3x}-e^{-3x}\right)$$

$$= \frac{9}{6}e^{3x}+\frac{9}{6}e^{-3x}+\frac{7}{6}e^{3x}-\frac{7}{6}e^{-3x}$$

$$= \frac{9}{6}e^{3x}+\frac{7}{6}e^{3x}+\frac{9}{6}e^{-3x}-\frac{7}{6}e^{-3x}$$

$$= \frac{16}{6}e^{3x}+\frac{2}{6}e^{-3x}$$

$$= \frac{8}{3}e^{3x}+\frac{1}{3}e^{-3x}$$

(237)

Chpt 15 The Inverse Laplace Transform Operator Y Linear

① evaluate

$$L^{-1}\left[\frac{p+2}{p^2+4}\right]$$

Ans

$$L^{-1}\left[\frac{p+2}{p^2+4}\right]$$

$$= L^{-1}\left[\frac{p}{p^2+4} + \frac{2}{p^2+4}\right]$$

$$= L^{-1}\left[\frac{p}{p^2+4}\right] + L^{-1}\left[\frac{2}{p^2+4}\right]$$

$$= L^{-1}\left[\frac{p}{p^2+2^2}\right] + L^{-1}\left[\frac{2}{p^2+2^2}\right]$$

$$= \cos 2x + \sin 2x$$

(238)

Chapt 15 The Inverse Laplace Transform Operation to Linear

8) evaluate

$$L^{-1}\left[\frac{p+5}{p^2+9}\right]$$

try

$$L^{-1}\left[\frac{p+5}{p^2+9}\right]$$

$$= L^{-1}\left[\frac{p}{p^2+9} + \frac{5}{p^2+9}\right]$$

$$= L^{-1}\left[\frac{p}{p^2+9}\right] + L^{-1}\left[\frac{5}{p^2+9}\right]$$

$$= L^{-1}\left[\frac{p}{p^2+3^2}\right] + 5\,L^{-1}\left[\frac{1}{p^2+3^2}\right]$$

$$= L^{-1}\left[\frac{p}{p^2+3^2}\right] + 5\,L^{-1}\left[\frac{\frac{3}{3}}{p^2+3^2}\right]$$

$$= L^{-1}\left[\frac{p}{p^2+3^2}\right] + \frac{5}{3}\,L^{-1}\left[\frac{3}{p^2+3^2}\right]$$

$$= \cos 3x + \frac{5}{3}\sin 3x$$

Chapt 15 The Inverse Laplace Transform Operator Volumes (239)

③ Evaluate

$$L^{-1}\left[\frac{3p-7}{p^2+25}\right]$$

Try

$$L^{-1}\left[\frac{3p-7}{p^2+25}\right]$$

$$= L^{-1}\left[\frac{3p}{p^2+25} - \frac{7}{p^2+25}\right]$$

$$= L^{-1}\left[\frac{3p}{p^2+25}\right] - L^{-1}\left[\frac{7}{p^2+25}\right]$$

$$= 3\,L^{-1}\left[\frac{p}{p^2+25}\right] - 7\,L^{-1}\left[\frac{1}{p^2+25}\right]$$

$$= 3\,L^{-1}\left[\frac{p}{p^2+5^2}\right] - 7\,L^{-1}\left[\frac{1}{p^2+5^2}\right]$$

$$= 3\,L^{-1}\left[\frac{p}{p^2+5^2}\right] - 7\,L^{-1}\left[\frac{\frac{5}{5}}{p^2+5^2}\right]$$

$$= 3\,L^{-1}\left[\frac{p}{p^2+5^2}\right] - \frac{7}{5}\,L^{-1}\left[\frac{5}{p^2+5^2}\right]$$

$$= 3\cos 5X - \frac{7}{5}\sin 5X$$

First-Order Homogeneous Linear D.E.'s
Chapt 16 With constant coefficients

Solving First-Order Homogeneous Linear D.E.'s
With constant Coefficients

① consider the first-order homogeneous linear D.E.
with constant coefficients

$$a \frac{dy}{dx} + by = 0$$

where a and b are constants

② we may solve such a D.E. by substituting
$y = C e^{mx}$ into the equation and
solving for m

First-Order Homogeneous Linear D.E.s With Constant Coefficients

$$a \frac{dy}{dx} + by = 0$$

let $y = C e^{mx}$

$y' = C e^{mx} \cdot m$

$y' = m C e^{mx}$

$a m C e^{mx} + b C e^{mx} = 0$

$C e^{mx} (am + b) = 0$

$am + b = 0$

$am = -b$

$m = -\frac{b}{a}$

$y = C e^{mx}$

$y = C e^{-\frac{b}{a}x}$

(242)   ∮ Intro

First-Order Homogeneous Linear D.E.'s

chapt 16   With Constant Coefficients

ex   solve the D.E.

$$3 \frac{dy}{dx} + 5y = 0$$

$$3 \frac{dy}{dx} + 5y = 0$$

let $y = Ce^{mx}$

$\quad y' = Ce^{mx} \cdot m$

$\quad y' = mCe^{mx}$

$$3mCe^{mx} + 5Ce^{mx} = 0$$

$$Ce^{mx}(3m+5) = 0$$

$$3m + 5 = 0$$

$$3m = -5$$

$$m = -\frac{5}{3}$$

$$y = Ce^{mx}$$

$$y = Ce^{-\frac{5}{3}x}$$

First-Order Homogeneous Linear D.E.

with Constant Coefficients

short cut method 1

$$3 \frac{dy}{dx} + 5y = 0$$

$$3m + 5 = 0$$

$$3m = -5$$

$$m = -\frac{5}{3}$$

$$y = Ce^{mx}$$

$$y = Ce^{-\frac{5}{3}x}$$

short cut method 2

$$3 \frac{dy}{dx} + 5y = 0$$

$$a \frac{dy}{dx} + by = 0$$

$$a = 3 \qquad b = 5$$

$$y = Ce^{-\frac{b}{a}x}$$

$$y = Ce^{-\frac{5}{3}x}$$

Intro

First-Order Homogeneous Linear D.E's
Chapt 16 With Constant Coefficients

check

$$3y' + 5y = 0$$

$$y = Ce^{-\frac{5}{3}x}$$

$$y' = Ce^{-\frac{5}{3}x}\left(-\frac{5}{3}\right)$$

$$y' = -\frac{5}{3}Ce^{-\frac{5}{3}x}$$

$$3\left(-\frac{5}{3}Ce^{-\frac{5}{3}x}\right) + 5\left(Ce^{-\frac{5}{3}x}\right) = 0$$

$$-5Ce^{-\frac{5}{3}x} + 5Ce^{-\frac{5}{3}x} = 0$$

$$0 = 0$$

VI

First-Order Homogeneous Linear D.E.'s & intro

Chapt 16 With Constant Coefficients

Ex solve the D.E.

$$7 \frac{dy}{dx} - 3y = 0$$

$$7 \frac{dy}{dx} - 3y = 0$$

$$\text{let } \quad y = Ce^{mx}$$

$$\frac{dy}{dx} = Ce^{mx} \cdot m$$

$$\frac{dy}{dx} = mCe^{mx}$$

$$7mCe^{mx} - 3Ce^{mx} = 0$$

$$Ce^{mx}(7m - 3) = 0$$

$$7m - 3 = 0$$

$$7m = 3$$

$$m = \frac{3}{7}$$

$$y = Ce^{mx}$$
$$y = Ce^{\frac{3}{7}x}$$

VII

(246)

First-Order Homogeneous Linear D.E.'s
Chapt 16   With Constant Coefficients

short cut method 1

$$7\frac{dy}{dx} - 3y = 0$$

$$7m - 3 = 0$$

$$7m = 3$$

$$m = \frac{3}{7}$$

$$y = Ce^{mx}$$
$$y = Ce^{\frac{3}{7}x}$$

short cut method 2

$$7\frac{dy}{dx} - 3y = 0$$

$$a\frac{dy}{dx} + by = 0$$

$$a = 7 \qquad b = -3$$

$$y = Ce^{-\frac{b}{a}x}$$
$$y = Ce^{-(\frac{-3}{7})x}$$
$$y = Ce^{\frac{3}{7}x}$$

First-Order Homogeneous Linear D.E.'s

Chapter 16 With Constant Coefficients

check

$$7 \frac{dy}{dx} - 3y = 0$$

$$y = Ce^{\frac{3}{7}x}$$
$$y' = Ce^{\frac{3}{7}x}\left(\frac{3}{7}\right)$$
$$y' = \frac{3}{7} Ce^{\frac{3}{7}x}$$

$$7\left(\frac{3}{7} Ce^{\frac{3}{7}x}\right) - 3 Ce^{\frac{3}{7}x} = 0$$

$$3 Ce^{\frac{3}{7}x} - 3 Ce^{\frac{3}{7}x} = 0$$

$$0 = 0$$

First-Order Homogeneous Linear D.E.'s

Chapt 16 With Constant Coefficients
Initial Value Problems

① an initial value problem includes
a differential equation and
one or more initial conditions

② the initial conditions change the
arbitrary constant or constants in the solution
to exact values

③④ an initial value problem involving
a first-order D.E. has one initial condition
$y(a) = b$ where a and b are numbers

④ an initial value problem involving
a second-order D.E. has two initial conditions
$y(a) = b$ where a and b are numbers
$y'(a) = c$ where a and c are numbers

note

① $y(a)$ and $y'(a)$ must have the same value of x

② if the values of x are different the problem
is called a boundary value problem

First-Order Homogeneous Linear D.E.'s

With Constant Coefficients

First-Order Initial Value Problems Involving
A First-Order Homogeneous Linear D.E.
With Constant Coefficients

① a first-order initial value problem involving
a first-order homogeneous linear D.E.
with constant coefficients   has the form

$$a \frac{dy}{dx} + by = 0$$

$$y(g) = h \qquad \text{where } g \text{ and } h \text{ are numbers}$$

② the solution of the D.E. is

$$y = Ce^{-\frac{b}{a}x}$$

by substituting $x = g$ and $y = h$
we may determine the value of C

③ the following example illustrates this technique

(250)

Intro.

First-Order Homogeneous Linear D.E.'s
Chapt 16 With Constant Coefficient

8x solve the init'l value problem (I.V.P.)

$$3 \frac{dy}{dx} - 2y = 0$$

$$y(0) = -5$$

$$3 \frac{dy}{dx} - 2y = 0$$

let $y = C e^{mx}$

$y' = C e^{mx} m$

$y' = m C e^{mx}$

$$3 m C e^{mx} - 2 C e^{mx} = 0$$

$$C e^{mx} (3m - 2) = 0$$

$$3m - 2 = 0$$

$$3m = 2$$

$$m = \frac{2}{3}$$

$$y = C e^{mx}$$

$$y = C e^{\frac{2}{3} x}$$

First-Order Homogeneous Linear D.E.s
Chapt 16 With Constant Coefficients

$$y(0) = -5 \qquad x = 0, \quad y = -5$$

$$y = Ce^{\frac{2}{3}x}$$

$$-5 = Ce^{0}$$

$$-5 = C(1)$$

$$-5 = C$$

$$C = -5$$

$$y = Ce^{\frac{2}{3}x}$$

$$y = -5\,e^{\frac{2}{3}x}$$

252

First-Order Homogeneous Linear D.E's

Chapt 16 With Constant Coefficients

check

$$3y' - 2y = 0$$

$$y = -5e^{\frac{2}{3}x}$$

$$y' = -5e^{\frac{2}{3}x}\left(\frac{2}{3}\right)$$

$$y' = -\frac{10}{3}e^{\frac{2}{3}x}$$

$$3\left(-\frac{10}{3}e^{\frac{2}{3}x}\right) - 2\left(-5e^{\frac{2}{3}x}\right) = 0$$

$$-10e^{\frac{2}{3}x} + 10e^{\frac{2}{3}x} = 0$$

$$0 = 0$$

$$y(0) = -5 \qquad\qquad x = 0, \ y = -5$$

$$y = -5e^{\frac{2}{3}x}$$

$$-5 = -5e^{0}$$

$$-5 = -5(1)$$

$$-5 = -5$$

Intro.

First-Order Homogeneous Linear D.E.'s
Chapt 16 With Constant Coefficients

Using The Laplace Transform To Solve
Initial Value Problems (I.V.P.'s)

① we may also use
the Laplace transform and
the inverse Laplace transform
to solve initial value problems

② the method is as follows
① take the Laplace transform of both sides
of the D.E.

② use
$$L[y'] = p L[y] - y(0)$$
note the initial condition $y(0)$ is included

③ solve for $L[y]$

④ to retrieve the function y
take the inverse transform of both sides
of the resulting equation
and use the relationship
$$L^{-1}[L[y]] = y$$

③ the following example illustrates
this technique

$\uparrow$ Intro

First-Order Homogeneous Linear D.E.'s

Chapter 16 With Constant Coefficient

ex use the Laplace transform to solve
the initial value problem (I.V.P.)

$$3 \frac{dy}{dx} - 2y = 0$$

$$y(0) = -5$$

$$3 \frac{dy}{dx} - 2y = 0$$

$$L\left[3\frac{dy}{dx} - 2y\right] = L[0]$$

$$L\left[3\frac{dy}{dx}\right] - L[2y] = 0$$

$$3 L\left[\frac{dy}{dx}\right] - 2 L[y] = 0$$

$$L[y'] = p\, L[y] - y(0)$$

$$3\{p\, L[y] - y(0)\} - 2 L[y] = 0$$

$$3\, p\, L[y] - 3\, y(0) - 2 L[y] = 0$$

$$3\, p\, L[y] - 3(-5) - 2 L[y] = 0$$

$$3\, p\, L[y] + 15 - 2 L[y] = 0$$

$$3\, p\, L[y] - 2 L[y] = -15$$

$$L[y]\,(3p - 2) = -15$$

$$L[y] = \frac{-15}{3p - 2}$$

$$L^{-1}\left[L[y]\right] = L^{-1}\left[\frac{-15}{3p - 2}\right]$$

Frist-Order Homogeneous Linear D.E.'s   Intro

Chapt 16   With Constant Coefficients

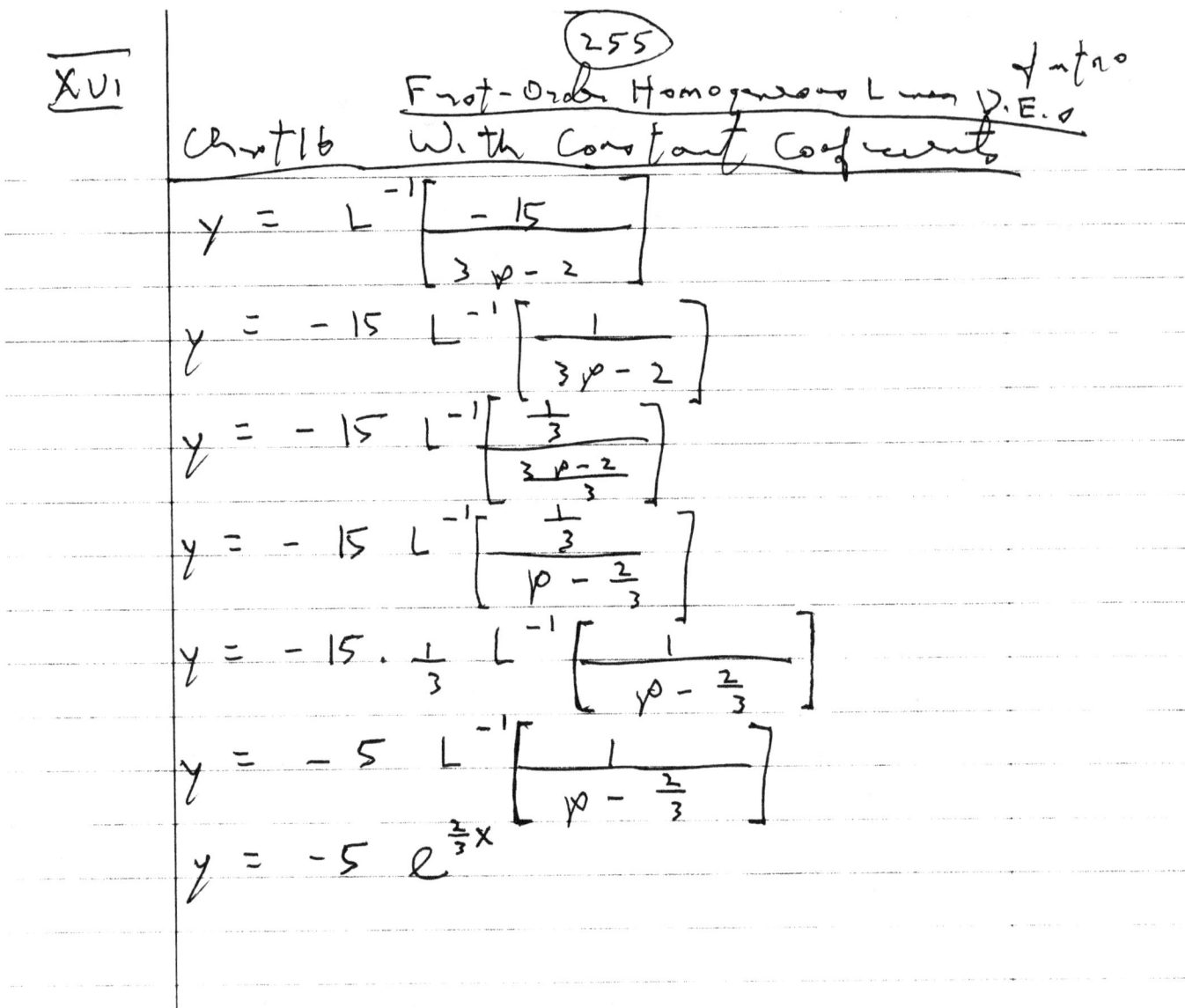

$$y = L^{-1}\left[\frac{-15}{3\rho - 2}\right]$$

$$y = -15 \; L^{-1}\left[\frac{1}{3\rho - 2}\right]$$

$$y = -15 \; L^{-1}\left[\frac{\frac{1}{3}}{\frac{3\rho - 2}{3}}\right]$$

$$y = -15 \; L^{-1}\left[\frac{\frac{1}{3}}{\rho - \frac{2}{3}}\right]$$

$$y = -15 \cdot \frac{1}{3} \; L^{-1}\left[\frac{1}{\rho - \frac{2}{3}}\right]$$

$$y = -5 \; L^{-1}\left[\frac{1}{\rho - \frac{2}{3}}\right]$$

$$y = -5 \; e^{\frac{2}{3}x}$$

(256)

First-Order Homogeneous Linear D.E.'s

chapt 16 With constant Coefficients

① solve the initial value problem (I.V.P.)
using two different methods
the conventional method   and
the Laplace transform method

$$\frac{dy}{dx} - y = 0$$

$$y(0) = 2$$

First-Order Homogeneous Linear D.E.'s

chapt 16   With constant coefficients
the conventional method

$$\frac{dy}{dx} - y = 0$$

let $y = Ce^{mx}$

$y' = Ce^{mx} \cdot m$

$y' = mCe^{mx}$

$mCe^{mx} - Ce^{mx} = 0$

$Ce^{mx}(m-1) = 0$

$m - 1 = 0$

$m = 1$

$y = Ce^{mx}$

$y = Ce^{x}$

$\underline{y(0) = 2}$ $x = 0, y = 2$

$y = Ce^{x}$

$2 = Ce^{0}$

$2 = C(1)$

$2 = C$

$C = 2$

$y = 2e^{x}$

IC

First-Order Homogeneous Linear D.E.'s

Chapt 16 With Constant Coefficients

(1 cont)

check

$$y' - y = 0$$

$$y = 2e^x$$
$$y' = 2e^x$$

$$2e^x - 2e^x = 0$$

$$0 = 0$$

$$y(0) = 2 \qquad x = 0, \ y = 2$$
$$y = 2e^x$$
$$2 = 2e^0$$
$$2 = 2(1)$$
$$2 = 2$$

First-Order Homogeneous Linear D.E.'s

chapter 16 With constant coefficients

the Laplace transform method

$$\frac{dy}{dx} - y = 0$$

$$y(0) = 2$$

$$\frac{dy}{dx} - y = 0$$

$$L\left[\frac{dy}{dx} - y\right] = L[0]$$

$$L\left[\frac{dy}{dx}\right] - L[y] = 0$$

$$L[y'] = pL[y] - y(0)$$

$$p\,L[y] - y(0) - L[y] = 0$$

$$p\,L[y] - 2 - L[y] = 0$$

$$p\,L[y] - L[y] = 2$$

$$L[y](p-1) = 2$$

$$L[y] = \frac{2}{p-1}$$

$$L^{-1}[L[y]] = L^{-1}\left[\frac{2}{p-1}\right]$$

$$y = 2L^{-1}\left[\frac{1}{p-1}\right]$$

$$y = 2e^x$$

(260)

First-Order Homogeneous Linear D.E.'s With Constant Coefficients

② Solve the initial value problem (I.V.P.) using two different methods the conventional method and the Laplace transform method

$$\frac{dy}{dx} - 2y = 0$$

$$y(0) = 5$$

First-Order Homogeneous Linear D.E's
Chapt 16 With Constant Coefficients
2 cont the conventional method

$$\frac{dy}{dx} - 2y = 0$$

let $y = Ce^{mx}$
$y' = Ce^{mx} \cdot m$
$y' = mCe^{mx}$

$mCe^{mx} - 2Ce^{mx} = 0$
$Ce^{mx}(m-2) = 0$
$m - 2 = 0$
$m = 2$

$y = Ce^{mx}$
$y = Ce^{2x}$

$y(0) = 5$ $x = 0, y = 5$
$y = Ce^{2x}$
$5 = Ce^{0}$
$5 = C(1)$
$5 = C$
$C = 5$

$y = 5e^{2x}$

(262)

First-Order Homogeneous Linear D.E.'s
with Constant Coefficients

(2 cont)

Check

$$y' - 2y = 0$$

$$y = 5e^{2x}$$

$$y' = 5e^{2x} \cdot 2$$

$$y' = 10e^{2x}$$

$$10e^{2x} - 2(5e^{2x}) = 0$$

$$10e^{2x} - 10e^{2x} = 0$$

$$0 = 0$$

$$\underline{y(0) = 5} \qquad x = 0, \ y = 5$$

$$y = 5e^{2x}$$

$$5 = 5e^{0}$$

$$5 = 5(1)$$

$$5 = 5$$

First-Order Homogeneous Linear D.E.'s
Cont'd With Constant Coefficients
using the Laplace transform method

$\frac{dy}{dx} - 2y = 0$

$y(0) = 5$

$\frac{dy}{dx} - 2y = 0$

$L\left[\frac{dy}{dx} - 2y\right] = L[0]$

$L\left[\frac{dy}{dx}\right] - L[2y] = 0$

$L\left[\frac{dy}{dx}\right] - 2L[y] = 0$

$L[y'] = pL[y] - y(0)$

$pL[y] - y(0) - 2L[y] = 0$

$pL[y] - 5 - 2L[y] = 0$

$pL[y] - 2L[y] = 5$

$L[y](p-2) = 5$

$L[y] = \frac{5}{p-2}$

$L^{-1}[L[y]] = L^{-1}\left[\frac{5}{p-2}\right]$

$y = 5L^{-1}\left[\frac{1}{p-2}\right]$

$y = 5e^{2x}$

First - Order Homogeneous Linear D.E.:
with Constant Coefficients

B) Solve the initial value problem (I.V.P.)
using two different methods
the conventional method   and
the Laplace transform method

$$\frac{dy}{dx} - 3y = 0$$

$$y(0) = -4$$

First-Order Homogeneous Linear D.E.'s
chratib With constant coefficient
(3 cont) the conventional method

$$\frac{dy}{dx} - 3y = 0$$

let $y = C e^{mx}$
$y' = C e^{mx} \cdot m$
$y' = m C e^{mx}$

$m C e^{mx} - 3 C e^{mx} = 0$
$C e^{mx} (m - 3) = 0$
$\quad\quad m - 3 = 0$
$\quad\quad\quad m = 3$

$y = C e^{mx}$
$y = C e^{3x}$

$\underline{y(0) = -4} \quad\quad x = 0, \; y = -4$
$y = C e^{3x}$
$-4 = C e^{0}$
$-4 = C(1)$
$-4 = C$
$C = -4$

$y = -4 e^{3x}$

First-Order Homogeneous Linear D.E.;
with constant coefficients

ch.#16

Check

$$y' - 3y = 0$$

$$y = -4 e^{3x}$$
$$y' = -4 e^{3x} (3)$$
$$y' = -12 e^{3x}$$

$$-12 e^{3x} - 3(-4 e^{3x}) = 0$$
$$-12 e^{3x} + 12 e^{3x} = 0$$
$$0 = 0$$

$$y(0) = -4 \qquad\qquad x = 0, \; y = -4$$
$$y = -4 e^{3x}$$
$$-4 = -4 e^{0}$$
$$-4 = -4 (1)$$
$$-4 = -4$$

First-Order Homogeneous Linear D.E.'s
With Constant Coefficients
the Laplace Transform method

$$\frac{dy}{dx} - 3y = 0$$

$$y(0) = -4$$

$$\frac{dy}{dx} - 3y = 0$$

$$L\left[\frac{dy}{dx} - 3y\right] = L[0]$$

$$L\left[\frac{dy}{dx}\right] - L[3y] = 0$$

$$L\left[\frac{dy}{dx}\right] - 3L[y] = 0$$

$$L[y'] = pL[y] - y(0)$$

$$pL[y] - y(0) - 3L[y] = 0$$

$$pL[y] - (-4) - 3L[y] = 0$$

$$pL[y] + 4 - 3L[y] = 0$$

$$pL[y] - 3L[y] = -4$$

$$L[y](p-3) = -4$$

$$L[y] = \frac{-4}{p-3}$$

$$L^{-1}[L[y]] = L^{-1}\left[\frac{-4}{p-3}\right]$$

$$y = -4 \, L^{-1}\left[\frac{1}{p-3}\right]$$

$$y = -4e^{3x}$$

First-Order Homogeneous Linear D.E.'s Chapt 16 With Constant Coefficients

(4) Solve the initial value problem (I.V.P.) using two different methods the conventional method and the Laplace transform method

$$\frac{dy}{dx} + y = 0$$

$$y(0) = 3$$

First-Order Homogeneous Linear D.E.'s

with Constant Coefficients

(4 int) the conventional method

$$\frac{dy}{dx} + y = 0$$

let $y = Ce^{mx}$
     $y' = Ce^{mx} \cdot m$
     $y' = mCe^{mx}$

$$mCe^{mx} + Ce^{mx} = 0$$
$$Ce^{mx}(m+1) = 0$$
$$m + 1 = 0$$
$$m = -1$$

$$y = Ce^{mx}$$
$$y = Ce^{-x}$$

$$y(0) = 3 \qquad\qquad x = 0, \; y = 3$$
$$y = Ce^{-x}$$
$$3 = Ce^{0}$$
$$3 = C(1)$$
$$3 = C$$
$$C = 3$$

$$y = 3e^{-x}$$

First-Order Homogeneous Linear D.E.'s
Chapt 16 With Constant Coefficients

(4 int) check

$$y' + y = 0$$

$$y_1 = 3e^{-x}$$
$$y' = 3e^{-x}(-1)$$
$$y' = -3e^{-x}$$

$$-3e^{-x} + 3e^{-x} = 0$$
$$0 = 0$$

$$\underline{y(0) = 3} \qquad x = 0, \; y = 3$$
$$y = 3e^{-x}$$
$$3 = 3e^{0}$$
$$3 = 3(1)$$
$$3 = 3$$

4 cont

27

First - Order Homogeneous Linear D.E.:
Chapt 16 With constant Coefficients
the Laplace transform method.

$$\frac{dy}{dx} + y = 0$$

$$y(0) = 3$$

$$\frac{dy}{dx} + y = 0$$

$$L\left[\frac{dy}{dx} + y\right] = L[0]$$

$$L\left[\frac{dy}{dx}\right] + L[y] = 0$$

$$L[y'] = p L[y] - y(0)$$

$$p L[y] - y(0) + L[y] = 0$$

$$p L[y] - 3 + L[y] = 0$$

$$p L[y] + L[y] = 3$$

$$L[y] (p+1) = 3$$

$$L[y] = \frac{3}{p+1}$$

$$L^{-1}[L[y]] = L^{-1}\left[\frac{3}{p+1}\right]$$

$$y = 3 L^{-1}\left[\frac{1}{p+1}\right]$$

$$y = 3 e^{-x}$$

First-Order Homogeneous Linear D.E.'s with Constant Coefficients

(5) solve the initial value problem (I.V.P.) using two different methods the convential method and the Laplace transform method

$$\frac{dy}{dx} + 2y = 0$$

$$y(0) = -2$$

First-Order Homogeneous Linear D.E.'s

chapt 16 With Constant Coefficients

the conventional method

$$\frac{dy}{dx} + 2y = 0$$

let $y = Ce^{mx}$

$y' = Ce^{mx} \cdot m$

$y' = mCe^{mx}$

$mCe^{mx} + 2Ce^{mx} = 0$

$Ce^{mx}(m+2) = 0$

$m + 2 = 0$

$m = -2$

$y = Ce^{mx}$

$y = Ce^{-2x}$

$\underline{y(0) = -2}$ 　　　　 $x = 0, \ y = -2$

$y = Ce^{-2x}$

$-2 = Ce^{0}$

$-2 = C(1)$

$-2 = C$

$C = -2$

$y = -2e^{-2x}$

First-Order Homogeneous Linear D.E.'s
Ch. #16 With Constant Coefficients

Check

$$y' + 2y = 0$$

$$y = -2e^{-2x}$$
$$y' = -2e^{-2x}(-2)$$
$$y' = 4e^{-2x}$$

$$4e^{-2x} + 2(-2e^{-2x}) = 0$$
$$4e^{-2x} - 4e^{-2x} = 0$$
$$0 = 0$$

$$\underline{y(0) = -2} \qquad x = 0, \; y = -2$$
$$y = -2e^{-2x}$$
$$-2 = -2e^{0}$$
$$-2 = -2(1)$$
$$-2 = -2$$

II D

P_{ro}

275

<u>First-Order Homogeneous Linear D.E.'s</u>
<u>Chapter 16 with constant coefficients</u>
<u>using the Laplace transform method</u>

$$\frac{dy}{dx} + 2y = 0$$

$$y(0) = -2$$

$$\frac{dy}{dx} + 2y = 0$$

$$L\left[\frac{dy}{dx} + 2y\right] = L[0]$$

$$L\left[\frac{dy}{dx}\right] + L[2y] = 0$$

$$L\left[\frac{dy}{dx}\right] + 2L[y] = 0$$

$$L[y'] = p\,L[y] - y(0)$$

$$p\,L[y] - y(0) + 2L[y] = 0$$
$$p\,L[y] - (-2) + 2L[y] = 0$$
$$p\,L[y] + 2 + 2L[y] = 0$$
$$p\,L[y] + 2L[y] = -2$$
$$L[y](p+2) = -2$$
$$L[y] = \frac{-2}{p+2}$$

$$L^{-1}[L[y]] = L^{-1}\left[\frac{-2}{p+2}\right]$$

$$y = -2\,L^{-1}\left[\frac{1}{p+2}\right]$$

$$y = -2e^{-2x}$$

VI A

Prob

(276)

Frist-Order Homogeneous Linear D.E.'s

Unit 16 With constant coefficients

(6) solve the initial value problem (I.V.P.)
using two different methods,
the conventional method   and
the Laplace transform method

$$\frac{dy}{dx} + 3y = 0$$

$$y(0) = -7$$

First-Order Homogeneous Linear D.E.'s
Chapt 16 With Constant Coefficient
(6 cont) the conventional method

$$\frac{dy}{dx} + 3y = 0$$

let $y = Ce^{mx}$
$y' = Ce^{mx} \cdot m$
$y' = mCe^{mx}$

$mCe^{mx} + 3Ce^{mx} = 0$
$Ce^{mx}(m+3) = 0$
$m+3 = 0$
$m = -3$

$y = Ce^{mx}$
$y = Ce^{-3x}$

$\underline{y(0) = -7} \qquad x = 0, \quad y = -7$
$y = Ce^{-3x}$
$-7 = Ce^{0}$
$-7 = C(1)$
$-7 = C$
$C = -7$

$y = -7e^{-3x}$

First-Order Homogeneous Linear D.E.:

Chapt 16, With Constant Coefficients

check

$$y' + 3y = 0$$

$$y = -7 e^{-3x}$$
$$y' = -7 e^{-3x} (-3)$$
$$y' = 21 e^{-3x}$$

$$21 e^{-3x} + 3(-7 e^{-3x}) = 0$$
$$21 e^{-3x} - 21 e^{-3x} = 0$$
$$0 = 0$$

$$\underline{y(0) = -7} \qquad x = 0, \; y = -7$$
$$y = -7 e^{-3x}$$
$$-7 = -7 e^{0}$$
$$-7 = -7 (1)$$
$$-7 = -7$$

Fist-Order Homogeneous Linear D.E.
Chapt 16 With constant Coefficents
6 cnt the Laplace transform method

$$\frac{dy}{dx} + 3y = 0$$

$$y(0) = -7$$

$$\frac{dy}{dx} + 3y = 0$$

$$L\left[\frac{dy}{dx} + 3y\right] = L[0]$$

$$L\left[\frac{dy}{dx}\right] + L[3y] = 0$$

$$L\left[\frac{dy}{dx}\right] + 3\,L[y] = 0$$

$$L[y'] = pL[y] - y(0)$$

$$p\,L[y] - y(0) + 3\,L[y] = 0$$

$$p\,L[y] - (-7) + 3\,L[y] = 0$$

$$p\,L[y] + 7 + 3\,L[y] = 0$$

$$p\,L[y] + 3\,L[y] = -7$$

$$L[y]\,(p+3) = -7$$

$$L[y] = \frac{-7}{p+3}$$

$$L^{-1}[L[y]] = L^{-1}\left[\frac{-7}{p+3}\right]$$

$$y = -7\,L^{-1}\left[\frac{1}{p+3}\right]$$

$$y = -7\,e^{-3x}$$

$\boxed{280}$

First-Order Nonhomogeneous Linear D.E.'s

chapt 17 with Constant Coefficients

Solving First-Order Nonhomogeneous Linear D.E.'s
with Constant Coefficients

① consider the first-order nonhomogeneous linear D.E.
with constant coefficients

$$a \frac{dy}{dx} + by = c(x)$$

where a and b are constants

② to solve such a D.E. we may use
either of two methods

ⓐ treating the D.E. as a
general first-order linear D.E. and
use the corresponding solution of such a D.E.

$$\frac{dy}{dx} + P(x)\,y = Q(x)$$

$$M = e^{\int P(x)\,dx}$$

$$My = \int M\,Q(x)\,dx \Bigg] \quad \text{the solution}$$

or

ⓑ use the method of undetermined coefficients

③ we shall concern ourselves with the first method
in this chapter

ⓐ we shall consider the method of undetermined coefficient
in a later chapter when we consider solving
second-order nonhomogeneous linear D.E.'s
with constant coefficients

④ the following examples illustrate
the first technique

Intro

First-Order Nonhomogeneous Linear D.E.'s

Chapt 17 With Constant Coefficients

ex   solve the D.E.

$$5 \frac{dy}{dx} + 3y = 7$$

$$5 \frac{dy}{dx} + 3y = 7$$

$$\frac{dy}{dx} + \frac{3}{5}y = \frac{7}{5}$$

$$\frac{dy}{dx} + P(x)y = Q(x)$$

$$P(x) = \frac{3}{5} \qquad Q(x) = \frac{7}{5}$$

$$\mu = e^{\int P \, dx}$$
$$\mu = e^{\int \frac{3}{5} \, dx}$$
$$\mu = e^{\frac{3}{5}x}$$

$$\mu y = \int \mu Q \, dx$$

$$e^{\frac{3}{5}x} y = \int e^{\frac{3}{5}x} \left(\frac{7}{5}\right) dx$$

$$e^{\frac{3}{5}x} y = \frac{7}{5} \int e^{\frac{3}{5}x} \, dx$$

$$e^{\frac{3}{5}x} y = \frac{7}{5} \left(\frac{5}{3} e^{\frac{3}{5}x} + c\right)$$

$$e^{\frac{3}{5}x} y = \frac{7}{3} e^{\frac{3}{5}x} + c$$

III

Chapt 17

First-Order Nonhomogeneous Linear D.E's Intro
With Constant Coefficients

$$\frac{e^{\frac{3}{5}x} y}{e^{\frac{3}{5}x}} = \frac{\frac{7}{3} e^{\frac{3}{5}x} + C}{e^{\frac{3}{5}x}}$$

$$y = \frac{7}{3} + C e^{-\frac{3}{5}x}$$

$$y = C e^{-\frac{3}{5}x} + \frac{7}{3}$$

First-Order Nonhomogeneous Linear D.E.: $+$ Intro

Chapt 17 With Constant Coefficients

check

$$5 \frac{dy}{dx} + 3y = 7$$

$$y = Ce^{-\frac{3}{5}x} + \frac{7}{3}$$

$$y' = Ce^{-\frac{3}{5}x}\left(-\frac{3}{5}\right) + 0$$

$$y' = -\frac{3}{5} Ce^{-\frac{3}{5}x}$$

$$5\left(-\frac{3}{5} Ce^{-\frac{3}{5}x}\right) + 3\left(Ce^{-\frac{3}{5}x} + \frac{7}{3}\right) = 7$$

$$-3Ce^{-\frac{3}{5}x} + 3Ce^{-\frac{3}{5}x} + 7 = 7$$

$$7 = 7$$

First-Order Nonhomogeneous Linear D.E.'s
Chapt 17 With Constant Coefficients

Q4 solve the D.E.

$$2 \frac{dy}{dx} - 3y = x$$

$$2 \frac{dy}{dx} - 3y = x$$

$$\frac{dy}{dx} - \frac{3}{2}y = \frac{x}{2}$$

$$\frac{dy}{dx} + P(x)y = Q(x)$$

$$P(x) = -\frac{3}{2} \qquad Q(x) = \frac{x}{2}$$

$$M = e^{\int P\,dx}$$

$$\mu = e^{\int (-\frac{3}{2})\,dx}$$

$$\mu = e^{-\frac{3}{2}x}$$

$$\mu y = \int \mu Q \, dx$$

$$e^{-\frac{3}{2}x} y = \int e^{-\frac{3}{2}x} \left(\frac{1}{2}x\right) dx$$

$$e^{-\frac{3}{2}x} y = \frac{1}{2} \int x e^{-\frac{3}{2}x} \, dx$$

$$\int x \, e^{-\frac{3}{2}x} \, dx$$

$$\int u \, dv = uv - \int v \, du$$

$$\text{let } u = x \qquad dv = e^{-\frac{3}{2}x} \, dx$$

$$du = dx \qquad \int dv = \int e^{-\frac{3}{2}x} \, dx$$

$$v = -\frac{2}{3} e^{-\frac{3}{2}x}$$

$$\int x \, e^{-\frac{3}{2}x} \, dx$$

$$= x \left(-\frac{2}{3} e^{-\frac{3}{2}x}\right) - \int \left(-\frac{2}{3} e^{-\frac{3}{2}x}\right) dx$$

$$= -\frac{2}{3} x \, e^{-\frac{3}{2}x} + \frac{2}{3} \int e^{-\frac{3}{2}x} \, dx$$

$$= -\frac{2}{3} x \, e^{-\frac{3}{2}x} + \frac{2}{3} \left[\left(-\frac{2}{3}\right) e^{-\frac{3}{2}x} + C\right]$$

$$= -\frac{2}{3} x \, e^{-\frac{3}{2}x} - \frac{4}{9} e^{-\frac{3}{2}x} + C$$

Chapt 17

First-Order Nonhomogeneous Linear D.E.'s

$+ m + n$

With constant Coefficients

$$e^{-\frac{3}{2}x} y = \frac{1}{2}\left(-\frac{2}{3}x e^{-\frac{3}{2}x} - \frac{4}{9} e^{-\frac{3}{2}x} + C\right)$$

$$e^{-\frac{3}{2}x} y = -\frac{1}{3} x e^{-\frac{3}{2}x} - \frac{2}{9} e^{-\frac{3}{2}x} + C$$

$$\frac{e^{-\frac{3}{2}x} y}{e^{-\frac{3}{2}x}} = \frac{-\frac{1}{3} x e^{-\frac{3}{2}x} - \frac{2}{9} e^{-\frac{3}{2}x} + C}{e^{-\frac{3}{2}x}}$$

$$y = -\frac{1}{3}x - \frac{2}{9} + C e^{\frac{3}{2}x}$$

$$y = C e^{\frac{3}{2}x} - \frac{1}{3}x - \frac{2}{9}$$

First-Order Nonhomogeneous Linear D.E.'s + Intro

Cont'17 With Constant Coefficients

check

$$2 \frac{dy}{dx} - 3y = x$$

$$y = Ce^{\frac{3}{2}x} - \frac{1}{3}x - \frac{2}{9}$$

$$y' = Ce^{\frac{3}{2}x}\left(\frac{3}{2}\right) - \frac{1}{3}(1) - 0$$

$$y' = \frac{3}{2}Ce^{\frac{3}{2}x} - \frac{1}{3}$$

$$2\left(\frac{3}{2}Ce^{\frac{3}{2}x} - \frac{1}{3}\right) - 3\left(Ce^{\frac{3}{2}x} - \frac{1}{3}x - \frac{2}{9}\right) = x$$

$$3Ce^{\frac{3}{2}x} - \frac{2}{3} - 3Ce^{\frac{3}{2}x} + x + \frac{2}{3} = x$$

$$x = x$$

Intro.

First-Order Nonhomogeneous Linear D.E's

Chapt 17 With Constant Coefficients

ex solve the initial value problem (I.V.P.)

$$5 \frac{dy}{dx} + 3y = e^x$$

$$y(0) = 4$$

$$5 \frac{dy}{dx} + 3y = e^x$$

$$\frac{dy}{dx} + \frac{3}{5}y = \frac{1}{5}e^x$$

$$\frac{dy}{dx} + P(x)y = Q(x)$$

$$P(x) = \frac{3}{5} \qquad Q(x) = \frac{1}{5}e^x$$

$$\mu = e^{\int P dx}$$

$$\mu = e^{\int \frac{3}{5} dx}$$

$$\mu = e^{\frac{3}{5}x}$$

$$\mu y = \int \mu Q \, dx$$

$$e^{\frac{3}{5}x} y = \int e^{\frac{3}{5}x} \left(\frac{1}{5} e^x\right) dx$$

$$e^{\frac{3}{5}x} y = \frac{1}{5} \int e^{\frac{8}{5}x} \, dx$$

First-Order Nonhomogeneous Linear D.E.
Chapter 17 With Constant Coefficients

$$e^{\frac{3}{5}x}\, y = \frac{1}{5}\left(\frac{5}{8}\, e^{\frac{8}{5}x} + C\right)$$

$$e^{\frac{3}{5}x}\, y = \frac{1}{8}\, e^{\frac{8}{5}x} + C$$

$$\frac{e^{\frac{3}{5}x}\, y}{e^{\frac{3}{5}x}} = \frac{\frac{1}{8}\, e^{\frac{8}{5}x} + C}{e^{\frac{3}{5}x}}$$

$$y = \frac{1}{8}\, e^{x} + C e^{-\frac{3}{5}x}$$

$$y = C e^{-\frac{3}{5}x} + \frac{1}{8}\, e^{x}$$

First-Order Nonhomogeneous Linear D.E.'s Intro

Chapt 17   With Constant Coefficients

check

$$5y' + 3y = e^x$$

$$y = Ce^{-\frac{3}{5}x} + \frac{1}{8}e^x$$

$$y' = Ce^{-\frac{3}{5}x}\left(-\frac{3}{5}\right) + \frac{1}{8}e^x$$

$$y' = -\frac{3}{5}Ce^{-\frac{3}{5}x} + \frac{1}{8}e^x$$

$$5\left(-\frac{3}{5}Ce^{-\frac{3}{5}x} + \frac{1}{8}e^x\right)$$

$$+3\left(Ce^{-\frac{3}{5}x} + \frac{1}{8}e^x\right)$$

$$= e^x$$

$$-3\cancel{Ce^{-\frac{3}{5}x}} + \frac{5}{8}e^x$$

$$+3\cancel{Ce^{-\frac{3}{5}x}} + \frac{3}{8}e^x$$

$$= e^x$$

$$\frac{5}{8}e^x + \frac{3}{8}e^x = e^x$$

$$\frac{8}{8}e^x = e^x$$

$$e^x = e^x$$

Chapter 17 — First-Order Nonhomogeneous Linear DE's Intro
With Constant Coefficients

$$y(0) = 4 \qquad x = 0, \ y = 4$$

$$y = Ce^{-\frac{3}{5}x} + \frac{1}{8}e^x$$

$$4 = Ce^0 + \frac{1}{8}e^0$$

$$4 = C(1) + \frac{1}{8}(1)$$

$$4 = C + \frac{1}{8}$$

$$C = 4 - \frac{1}{8}$$

$$C = \frac{32}{8} - \frac{1}{8}$$

$$C = \frac{31}{8}$$

$$y = Ce^{-\frac{3}{5}x} + \frac{1}{8}e^x$$

$$y = \frac{31}{8}e^{-\frac{3}{5}x} + \frac{1}{8}e^x$$

First-Order Nonhomogeneous Linear D.E.: †Intro

Chapter 17 With Constant Coefficients

&x use the Laplace transform to solve
the initial value problem (I. v. P.)

$$5 \frac{dy}{dx} + 3y = e^x$$

$$y(0) = 4$$

$$5 \frac{dy}{dx} + 3y = e^x$$

$$L\left[5 \frac{dy}{dx} + 3y\right] = L[e^x]$$

$$L\left[5 \frac{dy}{dx}\right] + L[3y] = L[e^x]$$

$$5 L\left[\frac{dy}{dx}\right] + 3 L[y] = L[e^x]$$

$$L[y'] = p L[y] - y(0)$$

$$5 \{p L[y] - y(0)\} + 3 L[y] = L[e^x]$$

$$5 p L[y] - 5 y(0) + 3 L[y] = L[e^x]$$

$$5 p L[y] - 5(4) + 3 L[y] = L[e^x]$$

$$5 p L[y] - 20 + 3 L[y] = L[e^x]$$

$$5 p L[y] + 3 L[y] = \frac{1}{p-1} + 20$$

$$L[y] (5p + 3) = \frac{1}{p-1} + 20$$

$$L[y] = \left(\frac{1}{5p+3}\right) \left(\frac{1}{p-1} + 20\right)$$

Chapt 17

(293)

Intro

First-Order Nonhomogeneous Linear D.E.'s
With Constant Coefficients

$$L[y] = \frac{1}{(5p+3)(p-1)} + \frac{20}{5p+3}$$

$$L[y] = \frac{1}{(p-1)(5p+3)} + \frac{20}{5p+3}$$

$$L^{-1}[L[y]] = L^{-1}\left[\frac{1}{(p-1)(5p+3)} + \frac{20}{5p+3}\right]$$

$$y = L^{-1}\left[\frac{1}{(p-1)(5p+3)}\right] + L^{-1}\left[\frac{20}{5p+3}\right]$$

(294)

First-Order Nonhomogeneous Linear D.E.'s Intro

Chapt 17   With Constant Coefficients

$$L^{-1}\left[\frac{1}{(p-1)(5p+3)}\right]$$

$$\frac{1}{(p-1)(5p+3)} = \frac{A}{p-1} + \frac{B}{5p+3}$$

$$= \frac{A}{p-1}\frac{5p+3}{5p+3} + \frac{B}{5p+3}\frac{p-1}{p-1}$$

$$= \frac{A(5p+3) + B(p-1)}{(p-1)(5p+3)}$$

$$1 = A(5p+3) + B(p-1)$$

$$A(5p+3) + B(p-1) = 1$$

$$5Ap + 3A + Bp - B = 1$$

$$5Ap + Bp + 3A - B = 1$$

$$(5A+B)p + (3A-B) = 1$$

$$5A + B = 0 \qquad 3A - B = 1$$

$$5A + B = 0$$
$$3A - B = 1$$

First-Order Nonhomogeneous Linear D.E's
Chapt 17 With Constant Coefficients ᵗⁿᵗʳᵒ

$$5A + B = 0$$
$$3A - B = 1$$

$$A = \frac{\begin{vmatrix} 0 & 1 \\ 1 & -1 \end{vmatrix}}{\begin{vmatrix} 5 & 1 \\ 3 & -1 \end{vmatrix}} = \frac{0-1}{-5-3} = \frac{-1}{-8} = \frac{1}{8}$$

$$B = \frac{\begin{vmatrix} 5 & 0 \\ 3 & 1 \end{vmatrix}}{\begin{vmatrix} 5 & 1 \\ 3 & -1 \end{vmatrix}} = \frac{5-0}{-5-3} = \frac{5}{-8} = -\frac{5}{8}$$

$$A = \frac{1}{8}, \quad B = -\frac{5}{8}$$

First-Order Nonhomogeneous Linear D.E.'s Intro
Chapt 17 With Constant Coefficients

$$L^{-1}\left[\frac{1}{(p-1)(5p+3)}\right]$$

$$= L^{-1}\left[\frac{\frac{1}{8}}{p-1} - \frac{\frac{5}{8}}{5p+3}\right]$$

$$= L^{-1}\left[\frac{\frac{1}{8}}{p-1}\right] - L^{-1}\left[\frac{\frac{5}{8}}{5p+3}\right]$$

$$= \frac{1}{8}L^{-1}\left[\frac{1}{p-1}\right] - \frac{5}{8}L^{-1}\left[\frac{1}{5p+3}\right]$$

$$= \frac{1}{8}e^{x} - \frac{5}{8}L^{-1}\left[\frac{\frac{1}{5}}{\frac{5p+3}{5}}\right]$$

$$= \frac{1}{8}e^{x} - \frac{5}{8}L^{-1}\left[\frac{\frac{1}{5}}{p+\frac{3}{5}}\right]$$

$$= \frac{1}{8}e^{x} - \frac{5}{8}\cdot\frac{1}{5}L^{-1}\left[\frac{1}{p+\frac{3}{5}}\right]$$

$$= \frac{1}{8}e^{x} - \frac{1}{8}L^{-1}\left[\frac{1}{p+\frac{3}{5}}\right]$$

$$= \frac{1}{8}e^{x} - \frac{1}{8}e^{-\frac{3}{5}x}$$

(297)

First-Order Nonhomogeneous Linear D.E's

Intro

Chapt 17 With Constant Coefficients

$$L^{-1}\left[\frac{20}{5p+3}\right]$$

$$= 20 \; L^{-1}\left[\frac{1}{5p+3}\right]$$

$$= 20 \; L^{-1}\left[\frac{\frac{1}{5}}{\frac{5p+3}{5}}\right]$$

$$= 20 \; L^{-1}\left[\frac{\frac{1}{5}}{p+\frac{3}{5}}\right]$$

$$= 20 \cdot \frac{1}{5} \; L^{-1}\left[\frac{1}{p+\frac{3}{5}}\right]$$

$$= 4 \; L^{-1}\left[\frac{1}{p+\frac{3}{5}}\right]$$

$$= 4 \; e^{-\frac{3}{5}x}$$

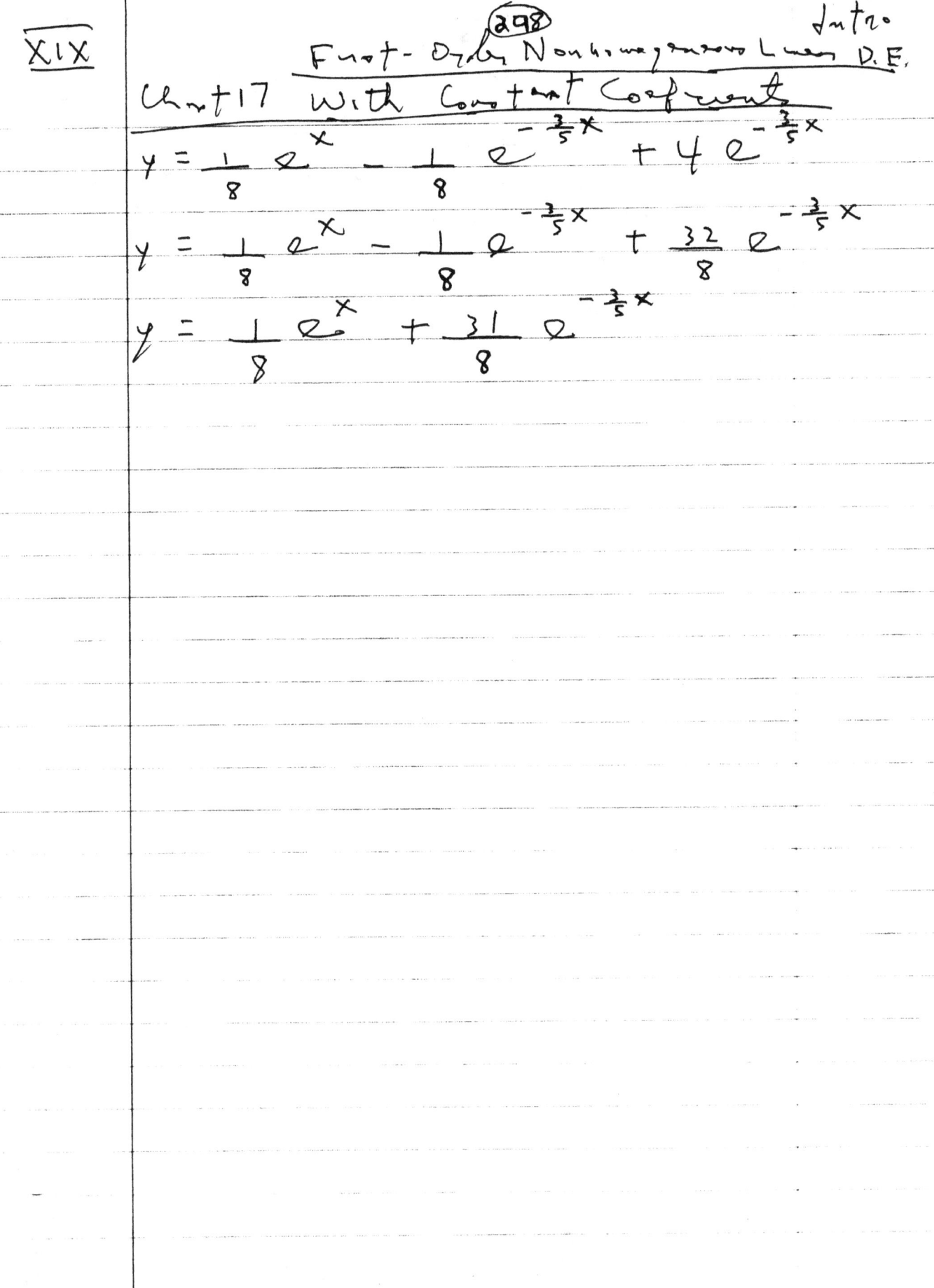

(298)

First-Order Nonhomogeneous Linear D.E.

Chart 17 With Constant Coefficients

$$y = \frac{1}{8} e^{x} - \frac{1}{8} e^{-\frac{3}{5}x} + 4 e^{-\frac{3}{5}x}$$

$$y = \frac{1}{8} e^{x} - \frac{1}{8} e^{-\frac{3}{5}x} + \frac{32}{8} e^{-\frac{3}{5}x}$$

$$y = \frac{1}{8} e^{x} + \frac{31}{8} e^{-\frac{3}{5}x}$$

First-Order Nonhomogeneous Linear D.E.'s

Chapt 17 with Constant Coefficients

① solve the initial value problem using
two different methods
the conventional method   and
the Laplace transform method

$$\frac{dy}{dx} + y = 5$$

$$y(0) = 7$$

First-Order Nonhomogeneous Linear D.E.

Chapt 17 With Constant Coefficients

(1 cont)

the conventional method

$$\frac{dy}{dx} + y = 5$$

$$\frac{dy}{dx} + P(x)y = Q(x)$$

$$P(x) = 1 \qquad Q(x) = 5$$

$$M = e^{\int P \, dx}$$

$$M = e^{\int 1 \cdot dx}$$

$$M = e^{x}$$

$$My = \int M Q \, dx$$

$$e^{x} y = \int e^{x} \cdot 5 \, dx$$

$$e^{x} y = 5 \int e^{x} \, dx$$

$$e^{x} y = 5 (e^{x} + C)$$

$$e^{x} y = 5 e^{x} + C$$

$$\frac{e^{x} y}{e^{x}} = \frac{5 e^{x} + C}{e^{x}}$$

$$y = 5 + C e^{-x}$$

$$y = C e^{-x} + 5$$

IC   Prob

First-Order Nonhomogeneous Linear D.E.

Chapt 17 With Constant Coefficients

(15.1f) $y(0) = 7$ $x = 0, y = 7$

$y = Ce^{-x} + 5$

$7 = Ce^{0} + 5$

$7 = C(1) + 5$

$7 = C + 5$

$c = 7 - 5$

$c = 2$

$y = Ce^{-x} + 5$

$y = 2e^{-x} + 5$

ID

(302)

First-Order Nonhomogeneous Linear D.E.'s

(1cont) Chapt 17. With constant Coefficient

check

$$y' + y = 5$$

$$y = 2e^{-x} + 5$$
$$y' = 2e^{-x}(-1) + 0$$
$$y' = -2e^{-x}$$

$$(-2e^{-x}) + (2e^{-x} + 5) = 5$$
$$-2e^{-x} + 2e^{-x} + 5 = 5$$
$$5 = 5$$

$$\underline{y(0) = 7} \qquad x = 0, y = 7$$
$$y = 2e^{-x} + 5$$
$$y = 2e^{0} + 5$$
$$y = 2(1) + 5$$
$$y = 2 + 5$$
$$y = 7$$

First-Order Nonhomogeneous Linear D.E.'s
Chapt 17 With Constant Coefficients

(16.1) using the Laplace transform method

$$y' + y = 5$$
$$y(0) = 7$$

$$y' + y = 5$$
$$L[y' + y] = L[5]$$
$$L[y'] + L[y] = L[5]$$
$$L[y'] = \rho L[y] - y(0)$$

$$\rho L[y] - y(0) + L[y] = L[5]$$
$$\rho L[y] - 7 + L[y] = L[5]$$
$$\rho L[y] + L[y] = L[5] + 7$$
$$L[y] (\rho + 1) = \frac{5}{\rho} + 7$$

$$L[y] = \left(\frac{1}{\rho + 1}\right) \left(\frac{5}{\rho} + 7\right)$$

$$L[y] = \frac{5}{\rho(\rho + 1)} + \frac{7}{\rho + 1}$$

$$L[y] = \frac{5}{\rho(\rho + 1)} + \frac{7\rho}{\rho(\rho + 1)}$$

$$L[y] = \frac{7\rho + 5}{\rho(\rho + 1)}$$

$$L^{-1}[L[y]] = L^{-1}\left[\frac{7\rho + 5}{\rho(\rho + 1)}\right]$$

$$y = L^{-1}\left[\frac{7\rho + 5}{\rho(\rho + 1)}\right]$$

First-Order Nonhomogeneous Linear D.E.'s

chapt 17 With constant coefficient

$$\frac{7p+5}{p(p+1)} = \frac{A}{p} + \frac{B}{p+1}$$

$$= \frac{A}{p} \cdot \frac{p+1}{p+1} + \frac{B}{p+1} \cdot \frac{p}{p}$$

$$= \frac{A(p+1) + Bp}{p(p+1)}$$

$$7p + 5 = A(p+1) + Bp$$
$$A(p+1) + Bp = 7p + 5$$
$$Ap + A + Bp = 7p + 5$$
$$Ap + Bp + A = 7p + 5$$
$$(A+B)p + A = 7p + 5$$
$$A + B = 7 \qquad A = 5$$
$$5 + B = 7$$
$$B = 7 - 5$$
$$B = 2$$

$$A = 5, \quad B = 2$$

$$\frac{A}{p} + \frac{B}{p+1}$$

$$= \frac{5}{p} + \frac{2}{p+1}$$

First-Order Nonhomogeneous Linear D.E.'s
With Constant Coefficients

(cont) $y = L^{-1}\left[\dfrac{7p+5}{p(p+1)}\right]$

$y = L^{-1}\left[\dfrac{5}{p} + \dfrac{2}{p+1}\right]$

$y = L^{-1}\left[\dfrac{5}{p}\right] + L^{-1}\left[\dfrac{2}{p+1}\right]$

$y = 5 L^{-1}\left[\dfrac{1}{p}\right] + 2 L^{-1}\left[\dfrac{1}{p+1}\right]$

$y = 5(1) + 2 e^{-x}$

$y = 5 + 2 e^{-x}$

$y = 2 e^{-x} + 5$

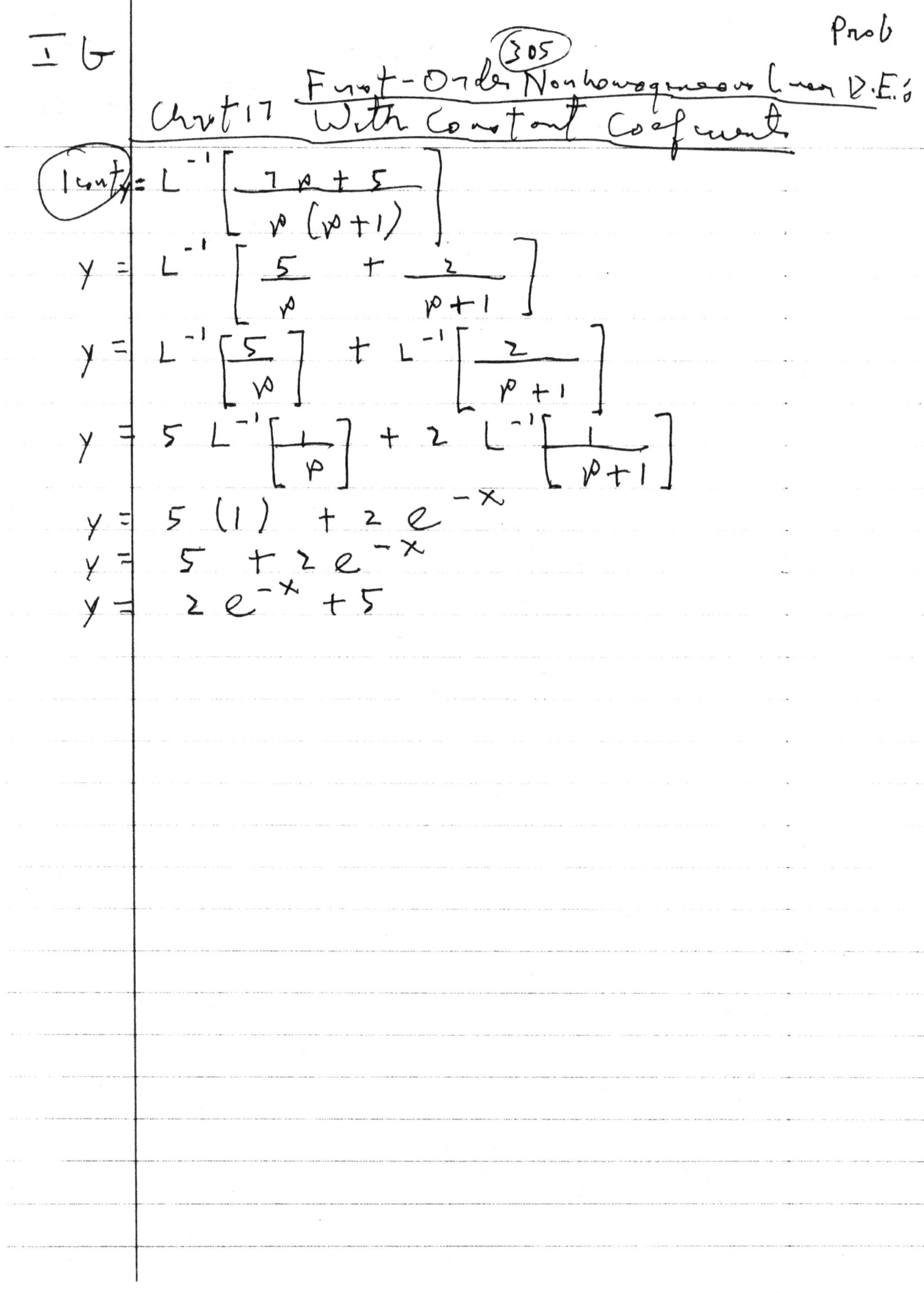

First-Order Nonhomogeneous Linear D.E.s

(306)

chapt 17 With constant Coefficients

② solve the initial value problem using two different methods the conventional method   and the Laplace transform method

$$\frac{dy}{dx} + y = x$$

$$y(0) = -5$$

First-Order Nonhomogeneous Linear D.E.:

Chapt 17 With Constant Coefficients

2cont the conventional method

$$\frac{dy}{dx} + y = x$$

$$\frac{dy}{dx} + P(x) y = Q(x)$$

$$P(x) = 1 \qquad Q(x) = x$$

$$\mu = e^{\int P dx}$$

$$\mu = e^{\int 1 \cdot dx}$$

$$\mu = e^{x}$$

$$\mu y = \int \mu Q \, dx$$

$$e^{x} y = \int e^{x} \cdot x \, dx$$

$$e^{x} y = \int x e^{x} \, dx$$

First-Order Nonhomogeneous Linear D.E.'s

Chapt 17   With Constant Coefficients

∑ cont

$$\int x e^x \, dx$$

$$\int u \, dv = uv - \int v \, du$$

$$\text{let } u = x \qquad dv = e^x \, dx$$

$$du = dx \qquad \int dv = \int e^x \, dx$$

$$v = e^x$$

$$\int x e^x \, dx$$

$$= x e^x - \int e^x \, dx$$

$$= x e^x - (e^x + C)$$

$$= x e^x - e^x + C$$

II. D

(309)

First-Order Nonhomogeneous Linear D.E.;

Chapt 17   With Constant Coefficients

(2 cont)

$$e^x y = x e^x - e^x + C$$

$$\frac{e^x y}{e^x} = \frac{x e^x - e^x + C}{e^x}$$

$$y = x - 1 + C e^{-x}$$

$$y = C e^{-x} + x - 1$$

(310)

First-Order Nonhomogeneous Linear D.E's

(cont) Chapt 17 With Constant Coefficient

$$y(0) = 7 \qquad x = 0, \quad y = 7$$

$$y = C e^{-x} + x - 1$$

$$7 = C e^{0} + 0 - 1$$

$$7 = C(1) - 1$$

$$7 = C - 1$$

$$C = 7 + 1$$

$$C = 8$$

$$y = 8 e^{-x} + x - 1$$

II F

First - Order Nonhomogeneous Linear D.E.'s

Chapt 17 With Constant Coefficients

(2 cont) check

$$y' + y = x$$

$$y = 8e^{-x} + x - 1$$
$$y' = 8e^{-x}(-1) + 1 - 0$$
$$y' = -8e^{-x} + 1$$

$$(-8e^{-x} + 1) + (8e^{-x} + x - 1) = x$$
$$-8e^{-x} + 1 + 8e^{-x} + x - 1 = x$$
$$x = x$$

$$\underline{y(0) = 7} \qquad x = 0, \; y = 7$$
$$y = 8e^{-x} + x - 1$$
$$y = 8e^{0} + 0 - 1$$
$$y = 8(1) - 1$$
$$y = 8 - 1$$
$$y = 7$$

First-Order Nonhomogeneous Linear DE.

Chapt 17 With Constant Coefficients

(2 cont) using the Laplace transform method

$$\frac{dy}{dx} + y = x$$

$$y(0) = 7$$

$$L\left[\frac{dy}{dx} + y\right] = L[x]$$

$$L\left[\frac{dy}{dx}\right] + L[y] = L[x]$$

$$L[y'] = \rho L[y] - y(0)$$

$$\rho L[y] - y(0) + L[y] = L[x]$$

$$\rho L[y] - 7 + L[y] = L[x]$$

$$\rho L[y] + L[y] = L[x] + 7$$

$$L[y](\rho + 1) = \frac{1!}{\rho^{1+1}} + 7$$

$$L[y] = \left(\frac{1}{\rho+1}\right)\left(\frac{1}{\rho^2} + 7\right)$$

$$L[y] = \frac{1}{\rho^2(\rho+1)} + \frac{7}{\rho+1}$$

$$L[y] = \frac{1}{\rho^2(\rho+1)} + \frac{7\rho^2}{\rho^2(\rho+1)}$$

$$L[y] = \frac{7\rho^2 + 1}{\rho^2(\rho+1)}$$

$$L^{-1}[L[y]] = L^{-1}\left[\frac{7\rho^2 + 1}{\rho^2(\rho+1)}\right]$$

First-Order Nonhomogeneous Linear D.E.'s

Chapt 17 with Constant Coefficients

(2 cont) $y = L^{-1} \left[\dfrac{7p^2 + 1}{p^2 (p+1)} \right]$

$$\frac{7p^2 + 1}{p^2 (p+1)} = \frac{A}{p} + \frac{B}{p^2} + \frac{C}{p+1}$$

$$= \frac{A}{p} \frac{p(p+1)}{p(p+1)} + \frac{B}{p^2} \frac{p+1}{p+1} + \frac{C}{p+1} \frac{p^2}{p^2}$$

$$= \frac{A \, p(p+1) + B(p+1) + C p^2}{p^2 (p+1)}$$

$7p^2 + 1 = A \, p(p+1) + B(p+1) + C p^2$

$A \, p(p+1) + B(p+1) + C p^2 = 7p^2 + 1$

$A (p^2 + p) + B(p+1) + C p^2 = 7p^2 + 1$

$A p^2 + A p + B p + B + C p^2 = 7p^2 + 1$

$A p^2 + C p^2 + A p + B p + B = 7p^2 + 1$

$(A+C) p^2 + (A+B) p + B = 7p^2 + 1$

$A + C = 7 \qquad A + B = 0 \qquad B = 1$

$-1 + C = 7 \qquad A + 1 = 0$

$C = 7 + 1 \qquad A = -1$

$C = 8$

$A = -1, \quad B = 1, \quad C = 8$

$$\frac{-1}{p} + \frac{1}{p^2} + \frac{8}{p+1}$$

314

Prob

First-Order Nonhomogeneous Linear D.E.'s

Chapt 17   With Constant Coeficients

2 Cont

$$y = L^{-1} \left[\frac{7p^2 + 1}{p^2(p+1)} \right]$$

$$y = L^{-1} \left[\frac{8}{p+1} + \frac{1}{p^2} - \frac{1}{p} \right]$$

$$y = L^{-1} \left[\frac{8}{p+1} \right] + L^{-1} \left[\frac{1}{p^2} \right] - L^{-1} \left[\frac{1}{p} \right]$$

$$y = 8 L^{-1} \left[\frac{1}{p+1} \right] + L^{-1} \left[\frac{1!}{p^{1+1}} \right] - L^{-1} \left[\frac{1}{p} \right]$$

$$y = 8 e^{-x} + x - 1$$

First - Order Nonhomogeneous Linear D.E.'s
Chapt 17 With Constant Coefficients

③ solve the initial value problem using
two different methods,
the conventional method   and
the Laplace transform method

$$\frac{dy}{dx} + y = e^{-x}$$

$$y(0) = -1$$

(316)

First-Order Nonhomogeneous Linear D.E.'s
Chapt 17 With Constant Coefficient
the conventional method

$$\frac{dy}{dx} + y = e^{-x}$$

$$\frac{dy}{dx} + P(x) y = Q(x)$$

$$P(x) = 1 \qquad Q(x) = e^{-x}$$

$$\mu = e^{\int P \, dx}$$

$$\mu = e^{\int 1 \cdot dx}$$

$$\mu = e^{x}$$

$$\mu y = \int \mu Q \, dx$$

$$e^{x} y = \int e^{x} \cdot e^{-x} \, dx$$

$$e^{x} y = \int e^{x-x} \, dx$$

$$e^{x} y = \int e^{0} \, dx$$

$$e^{x} y = \int 1 \cdot dx$$

$$e^{x} y = x + C$$

$$\frac{e^{x} y}{e^{x}} = \frac{x + C}{e^{x}}$$

$$y = x e^{-x} + C e^{-x}$$

$$y = C e^{-x} + x e^{-x}$$

First-Order Nonhomogeneous Linear D.E.'s

Chapt 17 With Constant Coefficients

3 cont) $y(0) = -1$ $x = 0, \ y = -1$

$$y = c e^{-x} + x e^{-x}$$

$$-1 = c e^{0} + 0$$

$$-1 = c(1)$$

$$-1 = c$$

$$c = -1$$

$$y = -e^{-x} + x e^{-x}$$

$$y = x e^{-x} - e^{-x}$$

First-Order Nonhomogeneous Linear D.E.:

<u>Chapt 17</u> With Constant Coefficients

(3 cont) <u>check</u>

$$y' + y = e^{-x}$$

$$y = -e^{-x} + x e^{-x}$$

$$y' = -e^{-x}(-1) + x e^{-x}(-1) + e^{-x}(1)$$

$$y' = e^{-x} - x e^{-x} + e^{-x}$$

$$y' = 2e^{-x} - x e^{-x}$$

$$(2e^{-x} - x e^{-x}) + (-e^{-x} + x e^{-x}) = e^{-x}$$

$$2e^{-x} - x e^{-x} - e^{-x} + x e^{-x} = e^{-x}$$

$$e^{-x} = e^{-x}$$

<u>$y(0) = -1$</u> $x = 0, \; y = -1$

$$y = -e^{-x} + x e^{-x}$$

$$y = -e^{0} + 0$$

$$y = -1$$

III E

3 cont

First-Order Nonhomogeneous Linear D.E.'s

Chapt 17 with constant coefficient

$\longrightarrow$ the Laplace transform method

$$\frac{dy}{dx} + y = e^{-x}$$

$$y(0) = -1$$

$$\frac{dy}{dx} + y = e^{-x}$$

$$L\left[\frac{dy}{dx} + y\right] = L[e^{-x}]$$

$$L\left[\frac{dy}{dx}\right] + L[y] = L[e^{-x}]$$

$$L[y'] = pL[y] - y(0)$$

$$p\,L[y] - y(0) + L[y] = L[e^{-x}]$$

$$p\,L[y] + 1 + L[y] = L[e^{-x}]$$

$$p\,L[y] + L[y] = L[e^{-x}] - 1$$

$$L[y](p+1) = \frac{1}{p+1} - 1$$

$$L[y] = \frac{1}{p+1}\left(\frac{1}{p+1} - 1\right)$$

$$L[y] = \frac{1}{(p+1)^2} - \frac{1}{p+1}$$

$$L[y] = \frac{1}{(p+1)^2} - \frac{p+1}{(p+1)^2}$$

$$L[y] = \frac{1 - p - 1}{(p+1)^2}$$

First-Order Nonhomogeneous Linear D.E.'s
Chapt 17 With Constant Coefficients

(3 cont) $L[y] = \dfrac{-p}{(p+1)^2}$

$L^{-1}[L[y]] = L^{-1}\left[\dfrac{-p}{(p+1)^2}\right]$

$y = L^{-1}\left[\dfrac{-p}{(p+1)^2}\right]$

$$\boxed{321}$$

First-Order Nonhomogeneous Linear D.E.s

Chapt 17   With Constant Coefficients

$$-\frac{p}{(p+1)^2} = \frac{A}{p+1} + \frac{B}{(p+1)^2}$$

$$= \frac{A}{p+1} \cdot \frac{p+1}{p+1} + \frac{B}{(p+1)^2}$$

$$= \frac{A(p+1) + B}{(p+1)^2}$$

$$-p = A(p+1) + B$$

$$A(p+1) + B = -p$$

$$Ap + A + B = -p$$

$$Ap + (A+B) = -p$$

$$A = -1 \qquad A + B = 0$$

$$-1 + B = 0$$

$$B = 1$$

$$A = -1, \quad B = 1$$

$$\frac{A}{p+1} + \frac{B}{(p+1)^2}$$

$$= \frac{-1}{p+1} + \frac{1}{(p+1)^2}$$

$$= \frac{1}{(p+1)^2} - \frac{1}{p+1}$$

First-Order Nonhomogeneous Linear D.E.'s
With Constant Coefficients

Chapter 17

(3 cont)

$$y = L^{-1}\left[\frac{-p}{(p+1)^2}\right]$$

$$y = L^{-1}\left[\frac{1}{(p+1)^2} - \frac{1}{p+1}\right]$$

$$y = L^{-1}\left[\frac{1}{(p+1)^2}\right] - L^{-1}\left[\frac{1}{p+1}\right]$$

$$y = L^{-1}\left[\frac{1!}{(p+1)^{1+1}}\right] - L^{-1}\left[\frac{1}{p+1}\right]$$

$$y = x e^{-x} - e^{-x}$$

First-Order Nonhomogeneous Linear D.E.'s

Chapt 17 With constant coefficient

① solve the D.E.

$$\frac{dy}{dx} + P(x) y = Q(x)$$

proof

$$\frac{dy}{dx} + P(x) y = Q(x)$$

let $u = e^{\int P(x) dx}$

multiply both sides of the D.E. by u

$$u \frac{dy}{dx} + u P(x) y = u Q(x)$$

$$u \frac{dy}{dx} + y [u P(x)] = u Q(x)$$

$$u \frac{dy}{dx} + y \frac{du}{dx} = u Q(x) \quad \leftarrow \text{this step}$$
$$\text{is magnified}$$
$$\text{on the next page}$$

$$\frac{d}{dx}(uy) = u Q(x)$$

$$\int \frac{d}{dx}(uy) dx = \int u Q(x) dx$$

$$uy = \int u Q(x) dx$$

I B

First - Order Nonhomogeneous Linear D.E.'s   Prob

Chapt 17   With Constant Coefficient

(16ut)

$$\mu = e^{\int p(x)\, dx}$$

$$\frac{d\mu}{dx} = e^{\int p(x)\, dx} \quad \frac{d}{dx}\left(\int p(x)\, dx\right)$$

$$\frac{d\mu}{dx} = e^{\int p(x)\, dx} \quad p(x)$$

pr-o/p

$$\frac{d\mu}{dx} = \mu\, p(x)$$

note

$$y = e^u$$

$$\frac{dy}{dx} = e^u\, \frac{du}{dx}$$

or

$$D_x\, e^u = e^u\, \frac{du}{dx}$$

I

(325)

Intro

Second-Order Homogeneous Linear D.E.'s

Chapt 18. With Constant Coefficients

Solving Second-Order Homogeneous Linear D.E.'s With Constant Coefficients

① consider the second-order homogeneous linear D.E.

$$a \frac{d^2 y}{dx^2} + b \frac{dy}{dx} + cy = 0$$

② we solve such a D.E. by substituting $y = Ce^{mx}$ into the D.E. and solving for m

③ we obtain two values of m and thus two different functions that satisfy the above D.E.

④ we lable each solution y_1 and y_2

⑤ finally we use the fact

if

y_1 is a solution to the D.E. and
y_2 is a solution to the D.E.

then

$y = y_1 + y_2$ is also a solution to the D.E.

⑥ this gives us the general solution of the above D.E. with two arbitrary constants in the solution

⑦ this is what we should expect for the general solution of a second-order D.E.

Second-Order Homogeneous Linear D.E.'s

Chapt 18 With Constant Coefficients

$$a \frac{d^2 y}{dx^2} + b \frac{dy}{dx} + cy = 0$$

let
$$y = C e^{mx}$$
$$y' = C e^{mx} \cdot m$$
$$y' = m C e^{mx}$$
$$y'' = m C e^{mx} \cdot m$$
$$y'' = m^2 C e^{mx}$$

$$a m^2 C e^{mx} + b m C e^{mx} + c C e^{mx} = 0$$
$$C e^{mx} (a m^2 + b m + c) = 0$$
$$a m^2 + b m + c = 0$$
$$m = \frac{-b \pm \sqrt{b^2 - 4ac}}{2a}$$

③ there are three cases to consider based upon
the value of the discriminant $b^2 - 4ac$

① if $b^2 - 4ac > 0$
the roots are real and distinct

② if $b^2 - 4ac = 0$
the roots are real and equal

③ if $b^2 - 4ac < 0$
the roots are conjugate complex numbers

Intro

Second-Order Homogeneous Linear D.E's

Chapt 18   With Constant Coefficients

(4) solutions

① the roots are real and unequal

the solution of the D.E. is given by

$$y = c_1 e^{m_1 x} + c_2 e^{m_2 x}$$

② the roots are real and equal

the solution of the D.E. is given by

$$y = c_1 e^{mx} + c_2 x e^{mx}$$

③ the roots are conjugate complex numbers.

$$m = r \pm si$$

the solution of the D.E. is given by

$$y = e^{rx} (c_1 \cos sx + c_2 \sin sx)$$

(5) note

① usually we use the notation $a \pm bi$
to designate conjugate complex numbers

② however since a and b are coefficients of the D.E.
we use the letters r and s instead

$r \pm si$ instead of

$a \pm bi$

③ however, in working actual problems
the $a \pm bi$ notation may be used

⑥ the following examples illustrate each case

(328)

Second-Order Homogeneous Linear D.E.'s Chapter 18 With Constant Coefficients

the roots are real and unequal

#x solve the D.E.

$$\frac{d^2y}{dx^2} - 3\frac{dy}{dx} + 2y = 0$$

$$y'' - 3y' + 2y = 0$$

$$m^2 - 3m + 2 = 0$$

$$(m-2)(m-1) = 0$$

$$m - 2 = 0 \qquad m - 1 = 0$$

$$m = 2 \qquad\qquad m = 1$$

$$m = 1, 2$$

$$y = C_1 e^{m_1 x} + C_2 e^{m_2 x}$$

$$y = C_1 e^{x} + C_2 e^{2x}$$

V

Second-Order Homogeneous Linear D.E.'s
Chapt 18 With Constant Coefficients
the roots are real and equal

#2 solve the D.E.

$$\frac{d^2 y}{dx^2} - 4 \frac{dy}{dx} + 4y = 0$$

$$y'' - 4y' + 4 = 0$$
$$m^2 - 4m + 4 = 0$$
$$(m-2)(m-2) = 0$$
$$m - 2 = 0 \qquad m - 2 = 0$$
$$m = 2 \qquad m = 2$$
$$m = 2, 2$$

$$y = C_1 e^{mx} + C_2 x e^{mx}$$
$$y = C_1 e^{2x} + C_2 x e^{2x}$$

Second-Order Homogeneous Linear D.E.'s
Chapt 18 With Constant Coefficients
the roots are conjugate complex numbers

Ex. solve the D.E.

$$\frac{d^2 y}{dx^2} + 2y = 0$$

$$y''_2 + 2y = 0$$
$$m^2 + 2 = 0$$
$$m^2 = -2$$
$$m = \pm \sqrt{-2}$$
$$m = \pm \sqrt{2(-1)}$$
$$m = \pm \sqrt{2} \, i$$

$$0 \pm \sqrt{2} \, i$$
$$a \pm b \, i$$
$$a = 0 \qquad b = \sqrt{2}$$

$$y = e^{ax}(c_1 \cos bx + c_2 \sin bx)$$
$$y = e^0 (c_1 \cos \sqrt{2} \, x + c_2 \sin \sqrt{2} \, x)$$
$$y = 1 (c_1 \cos \sqrt{2} \, x + c_2 \sin \sqrt{2} \, x)$$
$$y = c_1 \cos \sqrt{2} \, x + c_2 \sin \sqrt{2} \, x$$

Second-Order Homogeneous Linear D.E.'s
chapt 18 With Constant Coefficients
Initial Value Problem

① consider the second-order homogeneous linear D.E.
with constant coefficients

$$a \frac{d^2y}{dx^2} + b \frac{dy}{dx} + cy = 0$$

② because this is a second-order D.E.
there will be two initial conditions
in the D.E.'s initial value problem

$y(b) = g$ where b and g are numbers
$y'(b) = h$ where b and h are numbers

③ these two initial conditions give
exact values to the two arbitrary constants
c_1 and c_2 in the general solution

④ the following example illustrates this

VIII

(332)   Intro

Second-Order Homogeneous Linear D.E.'s
Chapt 18 With Constant Coefficients

ex   Solve the initial value Problem (I.V.P.)

$$\frac{d^2y}{dx^2} - 3\frac{dy}{dx} + 2y = 0$$

$$y(0) = 2$$
$$y'(0) = -1$$

$$y'' - 3y' + 2y = 0$$
$$m^2 - 3m + 2 = 0$$
$$(m-2)(m-1) = 0$$

$$m - 2 = 0 \qquad m - 1 = 0$$
$$m = 2 \qquad\quad m = 1$$
$$m = 1, 2$$

$$y = c_1 e^{m_1 x} + c_2 e^{m_2 x}$$
$$y = c_1 e^x + c_2 e^{2x}$$

Second-Order Homogeneous Linear D.E.'s

Chapt 18   With Constant Coefficients

$y(0) = 2$ $x = 0, y = 2$

$y = c_1 e^x + c_2 e^{2x}$

$2 = c_1 e^0 + c_2 e^0$

$2 = c_1 (1) + c_2 (1)$

$2 = c_1 + c_2$

$c_1 + c_2 = 2$

$y'(0) = -1$ $x = 0, y' = -1$

$y = c_1 e^x + c_2 e^{2x}$

$y' = c_1 e^x + 2 c_2 e^{2x}$

$-1 = c_1 e^0 + 2 c_2 e^0$

$-1 = c_1 (1) + 2 c_2 (1)$

$-1 = c_1 + 2 c_2$

$c_1 + 2 c_2 = -1$

$c_1 + c_2 = 2$

$c_1 + 2 c_2 = -1$

Second-Order Homogeneous Linear D.E.'s
Chapt 18 With Constant Coefficients.

$$c_1 + c_2 = 2$$
$$c_1 + 2c_2 = -1$$

$$c_1 = \frac{\begin{vmatrix} 2 & 1 \\ -1 & 2 \end{vmatrix}}{\begin{vmatrix} 1 & 1 \\ 1 & 2 \end{vmatrix}} = \frac{4+1}{2-1} = \frac{5}{1} = 5$$

$$c_2 = \frac{\begin{vmatrix} 1 & 2 \\ 1 & -1 \end{vmatrix}}{\begin{vmatrix} 1 & 1 \\ 1 & 2 \end{vmatrix}} = \frac{-1-2}{2-1} = \frac{-3}{1} = -3$$

$$y = c_1 e^x + c_2 e^{2x}$$
$$y = 5 e^x + (-3) e^{2x}$$
$$y = 5 e^x - 3 e^{2x}$$

Second-Order Homogeneous Linear D.E.'s
chapt 18 With constant Coeficients
Using The Laplace Transform To Solve
Initial Value Problems Involving
A Second-Order Homogeneous Linear D.E.
With Constant Coeficients

① we may also use
the Laplace transform and
the inverse Laplace transform
to solve the following type of
initial value problems (I.V.P.'s)

$$a \frac{d^2 y}{dx^2} + b \frac{dy}{dx} + cy = 0$$

$y(0) = e$ where e is a number

$y'(0) = f$ where f is a number

② we make use of the following formulas

$$L[y'] = p L[y] - y(0)$$

$$L[y''] = p^2 L[y] - p\, y(0) - y'(0)$$

§36   Intro

Second-Order Homogeneous Linear D.E.'s

Chapter 18 With constant Coefficients

Ex use the Laplace transform to solve
the initial value problem (I.V.P.)

$$\frac{d^2y}{dx^2} - 3\frac{dy}{dx} + 2y = 0$$

$$y(0) = 2$$
$$y'(0) = -1$$

$$\frac{d^2y}{dx^2} - 3\frac{dy}{dx} + 2y = 0$$

$$L\left[\frac{d^2y}{dx^2} - 3\frac{dy}{dx} + 2y\right] = L[0].$$

$$L\left[\frac{d^2y}{dx^2}\right] - L\left[3\frac{dy}{dx}\right] + L[2y] = 0$$

$$L\left[\frac{d^2y}{dx^2}\right] - 3L\left[\frac{dy}{dx}\right] + 2L[y] = 0$$

$$L[y'] = pL[y] - y(0)$$
$$L[y''] = p^2L[y] - py(0) - y'(0)$$

$$p^2L[y] - py(0) - y'(0) - 3\{pL[y] - y(0)\} + 2L[y] = 0$$
$$p^2L[y] - py(0) - y'(0) - 3pL[y] + 3y(0) + 2L[y] = 0$$
$$p^2L[y] - p(2) + 1 - 3pL[y] + 3(2) + 2L[y] = 0$$
$$p^2L[y] - 2p + 1 - 3pL[y] + 6 + 2L[y] = 0$$
$$p^2L[y] - 3pL[y] + 2L[y] = 2p - 1 - 6$$
$$p^2L[y] - 3pL[y] + 2L[y] = 2p - 7$$
$$L[y](p^2 - 3p + 2) = 2p - 7$$

(337)

Second-Order Homogeneous Linear D.E.'s

Chapt 18 With constant Coefficient

$$L[y] = \frac{2p - 7}{p^2 - 3p + 2}$$

$$L^{-1}[L[y]] = L^{-1}\left[\frac{2p - 7}{p^2 - 3p + 2}\right]$$

$$y = L^{-1}\left[\frac{2p - 7}{p^2 - 3p + 2}\right]$$

$$y = L^{-1}\left[\frac{2p - 7}{(p-1)(p-2)}\right]$$

$$\frac{2p - 7}{(p-1)(p-2)} = \frac{A}{p-1} + \frac{B}{p-2}$$

$$= \frac{A}{p-1}\frac{p-2}{p-2} + \frac{B}{p-2}\frac{p-1}{p-1}$$

$$= \frac{A(p-2) + B(p-1)}{(p-1)(p-2)}$$

$$2p - 7 = A(p-2) + B(p-1)$$
$$A(p-2) + B(p-1) = 2p - 7$$
$$Ap - 2A + Bp - B = 2p - 7$$
$$Ap + Bp - 2A - B = 2p - 7$$
$$(A + B)p + (-2A - B) = 2p - 7$$
$$A + B = 2 \qquad -2A - B = -7$$
$$2A + B = 7$$

$$A + B = 2$$
$$2A + B = 7$$

Second - Order Homogeneous Linear D.E.'s

Chapt 18 With Constant Coefficients

$$A + B = 2$$

$$2A + B = 7$$

$$A = \frac{\begin{vmatrix} 2 & 1 \\ 7 & 1 \end{vmatrix}}{\begin{vmatrix} 1 & 1 \\ 2 & 1 \end{vmatrix}} = \frac{2-7}{1-2} = \frac{-5}{-1} = 5$$

$$B = \frac{\begin{vmatrix} 1 & 2 \\ 2 & 7 \end{vmatrix}}{\begin{vmatrix} 1 & 1 \\ 2 & 1 \end{vmatrix}} = \frac{7-4}{1-2} = \frac{3}{-1} = -3$$

$$\frac{A}{p-1} + \frac{B}{p-2}$$

$$= \frac{5}{p-1} - \frac{3}{p-2}$$

Second-Order Homogeneous Linear D.E.
Chapt 18   With Constant Coefficients

$$y = L^{-1}\left[\frac{2p-7}{(p-1)(p-2)}\right]$$

$$y = L^{-1}\left[\frac{5}{p-1} - \frac{3}{p-2}\right]$$

$$y = L^{-1}\left[\frac{5}{p-1}\right] - L^{-1}\left[\frac{3}{p-2}\right]$$

$$y = 5\, L^{-1}\left[\frac{1}{p-1}\right] - 3\, L^{-1}\left[\frac{1}{p-2}\right]$$

$$y = 5\, e^{x} - 3\, e^{2x}$$

(340)

Second-Order Homogeneous Linear D.E.'s

Chapt 18 With constant coefficients

① solve the initial value problem
using two different methods
the convent and method and
the Laplace transform method

$$\frac{d^2 y}{dx^2} - 4y = 0$$

$$y(0) = 2$$
$$y'(0) = 5$$

I B

(341)

Second-Order Homogeneous Linear D.E.;

Chpt 18 With Constant Coefficient

(1 cont) the conventional method.

$$\frac{d^2 y}{dx^2} - 4y = 0$$

$$m^2 - 4 = 0$$

$$m^2 = 4$$

$$m = \pm\sqrt{4}$$

$$m = \pm 2$$

$$m = 2, -2$$

$$y = c_1 e^{m_1 x} + c_2 e^{m_2 x}$$

$$y = c_1 e^{2x} + c_2 e^{-2x}$$

IC

Second-Order Homogeneous Linear D.E.

Chapt 18 With Constant Coefficients

(1 cont)

$\underline{y(0) = 2}$ $X = 0, y = 2$

$y = C_1 e^{2x} + C_2 e^{-2x}$

$2 = C_1 e^0 + C_2 e^0$

$2 = C_1 (1) + C_2 (1)$

$2 = C_1 + C_2$

$C_1 + C_2 = 2$

$\underline{y'(0) = 5}$ $X = 0, y' = 5$

$y = C_1 e^{2x} + C_2 e^{-2x}$

$y' = C_1 e^{2x}(2) + C_2 e^{-2x}(-2)$

$y' = 2C_1 e^{2x} - 2C_2 e^{-2x}$

$5 = 2C_1 e^0 - 2C_2 e^0$

$5 = 2C_1 (1) - 2C_2 (1)$

$5 = 2C_1 - 2C_2$

$2C_1 - 2C_2 = 5$

$C_1 + C_2 = 2$

$2C_1 - 2C_2 = 5$

ID

(343)

Second - Order Homogeneous Linear D.E.'s
Chapt 18 With constant coefficients

(cont)

$C_1 + C_2 = 2$

$2C_1 - 2C_2 = 5$

$$C_1 = \frac{\begin{vmatrix} 2 & 1 \\ 5 & -2 \end{vmatrix}}{\begin{vmatrix} 1 & 1 \\ 2 & -2 \end{vmatrix}} = \frac{-4-5}{-2-2} = \frac{-9}{-4} = \frac{9}{4}$$

$$C_2 = \frac{\begin{vmatrix} 1 & 2 \\ 2 & 5 \end{vmatrix}}{\begin{vmatrix} 1 & 1 \\ 2 & -2 \end{vmatrix}} = \frac{5-4}{-2-2} = \frac{1}{-4} = -\frac{1}{4}$$

$C_1 = \frac{9}{4}, \quad C_2 = -\frac{1}{4}$

$y = C_1 e^{2x} + C_2 e^{-2x}$

$y = \frac{9}{4} e^{2x} - \frac{1}{4} e^{-2x}$

I E

Second-Order Homogeneous Linear D.E.'s

Chpt 18 With constant coefficients

(cont) check

$$y'' - 4y = 0$$

$$y = \frac{9}{4} e^{2x} - \frac{1}{4} e^{-2x}$$

$$y' = \frac{9}{4} e^{2x} (2) - \frac{1}{4} e^{-2x} (-2)$$

$$y' = \frac{9}{2} e^{2x} + \frac{1}{2} e^{-2x}$$

$$y'' = \frac{9}{2} e^{2x} (2) + \frac{1}{2} e^{-2x} (-2)$$

$$y'' = 9 e^{2x} - e^{-2x}$$

$$\left(9 e^{2x} - e^{-2x}\right) - 4\left(\frac{9}{4} e^{2x} - \frac{1}{4} e^{-2x}\right) = 0$$

$$9 e^{2x} - e^{-2x} - 9 e^{2x} + e^{-2x} = 0$$

$$0 = 0$$

$$y(0) = 2 \qquad\qquad x = 0, \; y = 2$$

$$y = \frac{9}{4} e^{2x} - \frac{1}{4} e^{-2x}$$

$$y = \frac{9}{4} e^{0} - \frac{1}{4} e^{0}$$

$$y = \frac{9}{4} (1) - \frac{1}{4} (1)$$

$$y = \frac{9}{4} - \frac{1}{4} = \frac{8}{4} = 2$$

$$y'(0) = 5 \qquad\qquad x = 0, \; y' = 5$$

$$y' = \frac{9}{2} e^{2x} + \frac{1}{2} e^{-2x}$$

$$y' = \frac{9}{2} e^{0} + \frac{1}{2} e^{0}$$

$$y' = \frac{9}{2} (1) + \frac{1}{2} (1)$$

$$y' = \frac{9}{2} + \frac{1}{2}$$

$$y' = \frac{10}{2}$$

$$y' = 5$$

I F

Prob

345

(1 cont)

Second-Order Homogeneous Linear D.E.
Chapt 18 With constant coefficients
the Laplace transform method

$$\frac{d^2y}{dx^2} - 4y = 0$$

$$y(0) = 2$$
$$y'(0) = 5$$

$$\frac{d^2y}{dx^2} - 4y = 0$$

$$L\left[\frac{d^2y}{dx^2} - 4y\right] = L[0]$$

$$L\left[\frac{d^2y}{dx^2}\right] - L[4y] = 0$$

$$L\left[\frac{d^2y}{dx^2}\right] - 4L[y] = 0$$

$$L[y'] = pL[y] - y(0)$$
$$L[y''] = p^2L[y] - py(0) - y'(0)$$

$$p^2L[y] - py(0) - y'(0) - 4L[y] = 0$$
$$p^2L[y] - p(2) - 5 - 4L[y] = 0$$
$$p^2L[y] - 2p - 5 - 4L[y] = 0$$
$$p^2L[y] - 4L[y] = 2p + 5$$
$$L[y](p^2 - 4) = 2p + 5$$
$$L[y] = \frac{2p + 5}{p^2 - 4}$$

$$L^{-1}[L[y]] = L^{-1}\left[\frac{2p + 5}{p^2 - 4}\right]$$

Chapt 18 Second-Order Homogeneous Linear D.E.'s with Constant Coefficients

(cont)

$$y = L^{-1}\left[\frac{2p+5}{p^2-4}\right]$$

$$y = L^{-1}\left[\frac{2p+5}{(p-2)(p+2)}\right]$$

Second Order Homogeneous Linear D.E.'s

Chapt 18   With Constant Coefficients

(cont) $\dfrac{2p + 5}{(p - 2)(p + 2)} = \dfrac{A}{p - 2} + \dfrac{B}{p + 2}$

$= \dfrac{A}{p-2} \cdot \dfrac{p+2}{p+2} + \dfrac{B}{p+2} \cdot \dfrac{p-2}{p-2}$

$= \dfrac{A(p+2) + B(p-2)}{(p-2)(p+2)}$

$2p + 5 = A(p+2) + B(p-2)$

$A(p+2) + B(p-2) = 2p + 5$

$Ap + 2A + Bp - 2B = 2p + 5$

$Ap + Bp + 2A - 2B = 2p + 5$

$(A + B)p + (2A - 2B) = 2p + 5$

$A + B = 2 \qquad 2A - 2B = 5$

$A + B = 2$

$2A - 2B = 5$

$A + B = 2$

$2A - 2B = 5$

$$A = \frac{\begin{vmatrix} 2 & 1 \\ 5 & -2 \end{vmatrix}}{\begin{vmatrix} 1 & 1 \\ 2 & -2 \end{vmatrix}} = \frac{-4-5}{-2-2} = \frac{-9}{-4} = \frac{9}{4}$$

$$B = \frac{\begin{vmatrix} 1 & 2 \\ 2 & 5 \end{vmatrix}}{\begin{vmatrix} 1 & 1 \\ 2 & -2 \end{vmatrix}} = \frac{5-4}{-2-2} = \frac{1}{-4} = -\frac{1}{4}$$

$$A = \frac{9}{4}, \quad B = -\frac{1}{4}$$

$$\frac{A}{p-2} + \frac{B}{p+2}$$

$$= \frac{\frac{9}{4}}{p-2} + \frac{-\frac{1}{4}}{p+2}$$

$$= \frac{\frac{9}{4}}{p-2} - \frac{\frac{1}{4}}{p+2}$$

Chapt 18   Second-Order Homogeneous Linear D.E.'s
With Constant Coefficients

(1 cont) $$y = L^{-1} \left[\frac{2p + 5}{(p-2)(p+2)} \right]$$

$$y = L^{-1} \left[\frac{\frac{9}{4}}{p-2} - \frac{\frac{1}{4}}{p+2} \right]$$

$$y = L^{-1} \left[\frac{\frac{9}{4}}{p-2} \right] - L^{-1} \left[\frac{\frac{1}{4}}{p+2} \right]$$

$$y = \frac{9}{4} L^{-1} \left[\frac{1}{p-2} \right] - \frac{1}{4} L^{-1} \left[\frac{1}{p+2} \right]$$

$$y = \frac{9}{4} e^{2x} - \frac{1}{4} e^{-2x}$$

$$\boxed{350}$$

Second-Order Homogeneous Linear D.E.'s

Chapt 18   with constant coefficients

① consider the D.E.

$$a\frac{d^2y}{dx^2} + b\frac{dy}{dx} + cy = 0$$

prove

~~proof~~

if

y_1 is a solution to the D.E.   and

y_2 is a solution to the D.E.

then

$y_T = y_1 + y_2$ is also a solution to the D.E.

Second-Order Homogeneous Linear D.E.'s with Constant Coefficients

(cont)

y_1 is a solution to the D.E.

$$a \frac{d^2 y}{dx^2} + b \frac{dy}{dx} + cy = 0$$

① $$a \frac{d^2 y_1}{dx^2} + b \frac{dy_1}{dx} + cy_1 = 0$$

Proof

y_2 is a solution to the D.E.

$$a \frac{d^2 y}{dx^2} + b \frac{dy}{dx} + cy = 0$$

② $$a \frac{d^2 y_2}{dx^2} + b \frac{dy_2}{dx} + cy_2 = 0$$

add equations ① and ②

I C

Second-Order Homogeneous Linear D.E.

Chapt 18 With Constant Coefficients

(1 cont)

$$a \frac{d^2 y_1}{dx^2} + b \frac{dy_1}{dx} + c y_1 = 0$$

$$a \frac{d^2 y_2}{dx^2} + b \frac{dy_2}{dx} + c y_2 = 0$$

Proof

$$a \frac{d^2 y_1}{dx^2} + a \frac{d^2 y_2}{dx^2}$$

$$+ b \frac{dy_1}{dx} + b \frac{dy_2}{dx}$$

$$+ c y_1 + c y_2$$

$$= 0$$

$$a \left(\frac{d^2 y_1}{dx^2} + \frac{d^2 y_2}{dx^2} \right)$$

$$+ b \left(\frac{dy_1}{dx} + \frac{dy_2}{dx} \right)$$

$$+ c (y_1 + y_2)$$

$$= 0$$

$$a \frac{d^2}{dx^2} (y_1 + y_2)$$

$$+ b \frac{d}{dx} (y_1 + y_2)$$

$$+ c (y_1 + y_2)$$

$$= 0$$

Second Order Homogeneous Linear D.E.

Chapt 18 With Constant Coefficients

② consider the D.E.

$$a \frac{d^2y}{dx^2} + b \frac{dy}{dx} + cy = 0$$

where a, b, and c are constants

Proof: substituting $y = ce^{mx}$ into the D.E results in the auxilary equation

$$am^2 + bm + c = 0$$

where

$$m = \frac{-b \pm \sqrt{b^2 - 4ac}}{2a}$$

prove each of the following solutions corresponding to each of the three possible cases for the discriminent

case 1

if $b^2 - 4ac > 0$

then $y = c_1 e^{m_1 x} + c_2 e^{m_2 x}$

case 2

if $b^2 - 4ac = 0$

then $y = c_1 e^{mx} + c_2 x e^{mx}$

case 3

if $b^2 - 4ac < 0$

then $y = e^{rx}(c_1 \cos sx + c_2 \sin sx)$

Chpt 18 — Second Order Homogeneous Linear DEs
With Constant Coefficients

2 cont

$$a\frac{d^2y}{dx^2} + b\frac{dy}{dx} + cy = 0$$

let $y = Ce^{mx}$ ✓

proof

$$y' = Ce^{mx} \cdot m$$

$$y' = mCe^{mx}$$ ✓

$$y'' = mCe^{mx} \cdot m$$

$$y'' = m^2Ce^{mx}$$ ✓

$$a(m^2Ce^{mx}) + b(mCe^{mx}) + c(Ce^{mx}) = 0$$

$$am^2Ce^{mx} + bmCe^{mx} + cCe^{mx} = 0$$

$$Ce^{mx}(am^2 + bm + c) = 0$$

$$am^2 + bm + c = 0$$

$$m = \frac{-b \pm \sqrt{b^2 - 4ac}}{2a}$$

$$m_1 = \frac{-b + \sqrt{b^2 - 4ac}}{2a}$$

$$m_2 = \frac{-b - \sqrt{b^2 - 4ac}}{2a}$$

Chapt 18 $\underline{\text{Second Order Homogeneous Linear D.E.}}$
$\underline{\text{With Constant Coefficients}}$

2 cont $\underline{\text{case 1}}$ $b^2 - 4ac > 0$

$$y = C e^{mx}$$

$$y_1 = C_1 e^{m_1 x}$$

prove where $m_1 = \dfrac{-b + \sqrt{b^2 - 4ac}}{2a}$

$$y_2 = C_2 e^{m_2 x}$$

where $m_2 = \dfrac{-b - \sqrt{b^2 - 4ac}}{2a}$

since both y_1 and y_2 satisfy the DE

so does $y = y_1 + y_2$ or

$$y = C_1 e^{m_1 x} + C_2 e^{m_2 x}$$

Second Order Homogeneous Linear DEs
Chapt 18 With Constant Coefficients

$\sum$ cont

case 2 $\qquad b^2 - 4ac = 0$

$$a \frac{d^2 y}{dx^2} + b \frac{dy}{dx} + cy = 0$$

$$m = \frac{-b \pm \sqrt{b^2 - 4ac}}{2a}$$

proofs

$$b^2 - 4ac = 0$$

$$m = \frac{-b \pm 0}{2a}$$

$$m = \frac{-b}{2a}$$

$$m = -\frac{b}{2a}$$

$$y_1 = C_1 e^{mx}$$
$$y_1 = C_1 e^{-\frac{b}{2a} x}$$

let
$$y_2 = C_2 x e^{-\frac{b}{2a} x}$$

we must now show that each y_1 and y_2 satisfy the DE

Second Order Homogeneous Linear DEs
Chpt 18 With Constant Coefficients

(2 cont)

$$y_1 = C_1 e^{-\frac{b}{2a}x}$$

$$a\frac{d^2y}{dx^2} + b\frac{dy}{dx} + cy = 0$$

Proof

$$y = C_1 e^{-\frac{b}{2a}x}$$

$$y' = C_1 e^{-\frac{b}{2a}x}\left(-\frac{b}{2a}\right)$$

$$y'' = C_1 e^{-\frac{b}{2a}x}\frac{b^2}{4a^2}$$

$$a\, C_1 e^{-\frac{b}{2a}x}\frac{b^2}{4a^2} + b\left(-C_1 e^{-\frac{b}{2a}x}\frac{b}{2a}\right) + c\, C_1 e^{-\frac{b}{2a}x} = 0$$

$$a C_1 e^{-\frac{b}{2a}x}\frac{b^2}{4a^2} - C_1 e^{-\frac{b}{2a}x}\frac{b^2}{2a} + c\, C_1 e^{-\frac{b}{2a}x} = 0$$

$$C_1 e^{-\frac{b}{2a}x}\left(\frac{ab^2}{4a^2} - \frac{b^2}{2a} + c\right) = 0$$

$$C_1 e^{-\frac{b}{2a}x}\left(\frac{b^2}{4a} - \frac{b^2}{2a} + c\right) = 0$$

$$\frac{b^2}{4a} - \frac{b^2}{2a} + c = 0$$

$$\frac{b^2}{4a} - \frac{b^2}{2a}\cdot\frac{2}{2} + c\frac{4a}{4a} = 0$$

$$\frac{b^2 - 2b^2 + 4ac}{4a} = 0$$

$$\frac{-b^2 + 4ac}{4a} = 0$$

$$-b^2 + 4ac = 0$$

$$b^2 - 4ac = 0$$

Chapt 18 Second Order Homogeneous Linear D.E. With Constant Coefficients

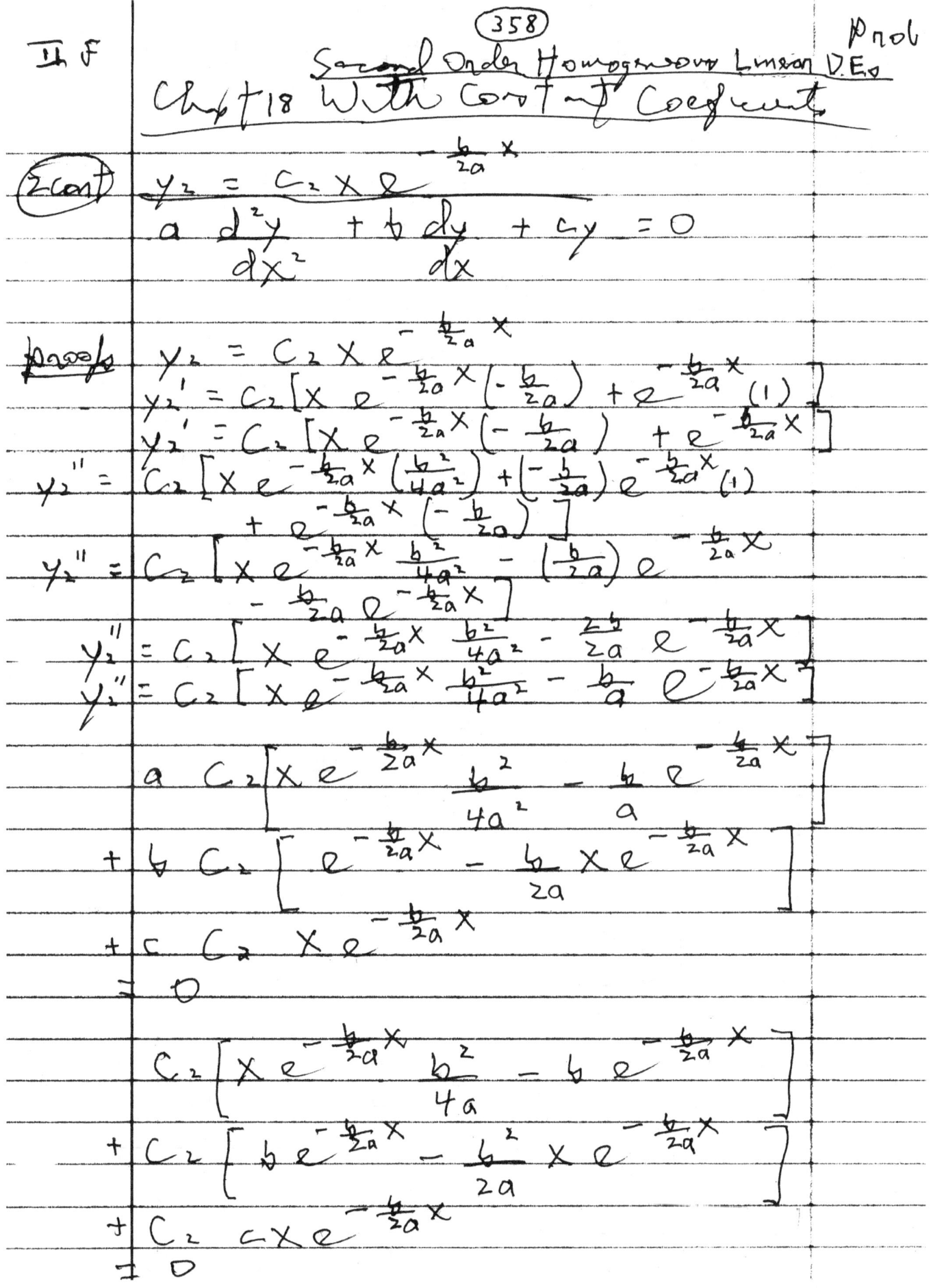

(2 cont)

$$y_2 = C_2 x e^{-\frac{b}{2a}x}$$

$$a\frac{d^2y}{dx^2} + b\frac{dy}{dx} + cy = 0$$

Proof

$$y_2 = C_2 x e^{-\frac{b}{2a}x}$$

$$y_2' = C_2\left[x e^{-\frac{b}{2a}x}\left(-\frac{b}{2a}\right) + e^{-\frac{b}{2a}x}(1)\right]$$

$$y_2' = C_2\left[x e^{-\frac{b}{2a}x}\left(-\frac{b}{2a}\right) + e^{-\frac{b}{2a}x}\right]$$

$$y_2'' = C_2\left[x e^{-\frac{b}{2a}x}\left(\frac{b^2}{4a^2}\right) + \left(-\frac{b}{2a}\right)e^{-\frac{b}{2a}x}(1) + e^{-\frac{b}{2a}x}\left(-\frac{b}{2a}\right)\right]$$

$$y_2'' = C_2\left[x e^{-\frac{b}{2a}x}\frac{b^2}{4a^2} - \left(\frac{b}{2a}\right)e^{-\frac{b}{2a}x} - \frac{b}{2a}e^{-\frac{b}{2a}x}\right]$$

$$y_2'' = C_2\left[x e^{-\frac{b}{2a}x}\frac{b^2}{4a^2} - \frac{2b}{2a}e^{-\frac{b}{2a}x}\right]$$

$$y_2'' = C_2\left[x e^{-\frac{b}{2a}x}\frac{b^2}{4a^2} - \frac{b}{a}e^{-\frac{b}{2a}x}\right]$$

$$a C_2\left[x e^{-\frac{b}{2a}x}\frac{b^2}{4a^2} - \frac{b}{a}e^{-\frac{b}{2a}x}\right]$$

$$+ b C_2\left[e^{-\frac{b}{2a}x} - \frac{b}{2a}x e^{-\frac{b}{2a}x}\right]$$

$$+ c C_2 x e^{-\frac{b}{2a}x}$$

$$= 0$$

$$C_2\left[x e^{-\frac{b}{2a}x}\frac{b^2}{4a} - b e^{-\frac{b}{2a}x}\right]$$

$$+ C_2\left[b e^{-\frac{b}{2a}x} - \frac{b^2}{2a}x e^{-\frac{b}{2a}x}\right]$$

$$+ C_2 c x e^{-\frac{b}{2a}x}$$

$$= 0$$

Chapt 18   With Constant Coefficient

$\left(C_2 e^{-\frac{b}{2a}x}\right)$ (cont)

$$\left(x\frac{b^2}{4a} - b + b - \frac{b^2}{2a}x + cx\right)$$

$$= 0$$

proof

$$x\frac{b^2}{4a} - b + b - \frac{b^2}{2a}x + cx = 0$$

$$x\frac{b^2}{4a} - \frac{b^2}{2a}x + cx = 0$$

$$x\left(\frac{b^2}{4a} - \frac{b^2}{2a} + c\right) = 0$$

$$\frac{b^2}{4a} - \frac{b^2}{2a} + c = 0$$

$$\frac{b^2}{4a} - \frac{b^2}{2a}\frac{2}{2} + c\frac{4a}{4a} = 0$$

$$\frac{b^2}{4a} - \frac{2b^2}{4a} + \frac{4ac}{4a} = 0$$

$$\frac{b^2 - 2b^2 + 4ac}{4a} = 0$$

$$\frac{-b^2 + 4ac}{4a} = 0$$

$$-b^2 + 4ac = 0$$

$$b^2 - 4ac = 0$$

Second Order Homogeneous Linear D.E.

Unit 18   With Constant Coefficients

(2 cont)   therefore

since

$$y_1 = C_1 e^{mx} \qquad \text{where } m = -\frac{b}{2a}$$

is a solution of the D.E

proof $a \dfrac{d^2 y}{dx^2} + b \dfrac{dy}{dx} + cy = 0$

and

$$y_2 = C_2 x e^{mx} \qquad \text{where } m = -\frac{b}{2a}$$

is a solution of the D.E

we have

$$y = y_1 + y_2$$

or

$$y = C_1 e^{mx} + C_2 x e^{mx}$$

is the general solution of

$$a \dfrac{d^2 y}{dx^2} + b \dfrac{dy}{dx} + cy = 0$$

where

$$b^2 - 4ac = 0$$

Second Order Homogeneous Linear D E
Chapt 18 With Constant Coefficients

(∑ cont) Case 3 $b^2 - 4ac < 0$

$$a\frac{d^2y}{dx^2} + b\frac{dy}{dx} + cy = 0$$

$$m = \frac{-b \pm \sqrt{b^2 - 4ac}}{2a}$$

proof

$b^2 - 4ac < 0$

therefore

$m = r \pm si$

$m_1 = r + si$

$m_2 = r - si$

$y_1 = C_1 e^{m_1 x}$

$y_1 = C_1 e^{(r+si)x}$

$y_1 = C_1 e^{rx + sxi}$

$y_1 = C_1 e^{rx} \cdot e^{sxi}$ ✓

$y_2 = C_2 e^{m_2 x}$

$y_2 = C_2 e^{(r-si)x}$

$y_2 = C_2 e^{rx - sxi}$

$y_2 = C_2 e^{rx} \cdot e^{-sxi}$ ✓

Second Order Homogeneous Linear DE.

Unit 18 With Constant Coefficients

(2con)

$$e^{\theta i} = \cos\theta + (\sin\theta) i$$

$$e^{sxi} = \cos(sx) + [\sin(sx)] i$$

$$e^{-\theta i} = \cos\theta - (\sin\theta) i$$

$$e^{-sxi} = \cos(sx) - [\sin(sx)] i$$

$$y_T = y_1 + y_2$$

$$y_T = C_1 e^{nx} \left\{ \cos(sx) + [\sin(sx)] i \right\}$$
$$+ C_2 e^{nx} \left\{ \cos(sx) - [\sin(sx)] i \right\}$$

$$y_T = e^{rx} \left\{ C_1 \cos(sx) + C_1 [\sin(sx)] i \right\}$$
$$+ e^{rx} \left\{ C_2 \cos(sx) - C_2 [\sin(sx)] i \right\}$$

$$y_T = e^{rx} \left\{ C_1 \cos(sx) + C_1 [\sin(sx)] i \right.$$
$$\left. + C_2 \cos(sx) - C_2 [\sin(sx)] i \right\}$$

$$y_T = e^{rx} \left\{ C_1 \cos(sx) + C_2 \cos(sx) \right.$$
$$\left. + C_1 [\sin(sx)] i - C_2 [\sin(sx)] i \right\}$$

$$y_T = e^{rx} \left[(C_1 + C_2) \cos(sx) \right.$$
$$\left. + (C_1 - C_2) \sin(sx) i \right]$$

$$y_T = e^{rx} \left[(C_1 + C_2) \cos(sx) \right.$$
$$\left. + (C_1 - C_2) i \sin(sx) \right]$$

Chapt 18   Second Order Homogeneous Linear DE's With Constant Coefficients

(2 cont)

let $\quad C^* = C_1 + C_2$

$\qquad C^{**} = (C_1 - C_2) i$

Proof

$$y_T = e^{rx} [C^* \cos(sx) + C^{**} \sin(sx)]$$

using more conventional notation

$$y_T = e^{rx} [C_1 \cos(sx) + C_2 \sin(sx)]$$

note

$i = \sqrt{-1}$ which is a legitimate number

Second-Order Nonhomogeneous Linear D.E.'s
Chapt 19 With Constant Coefficients

Solving Second-Order Nonhomogeneous Linear D.E.'s
With Constant Coefficients

① consider the second-order nonhomogeneous
linear D.E. with constant coefficient

$$a \frac{d^2y}{dx^2} + b \frac{dy}{dx} + cy = d(x)$$

and its corresponding homogeneous D.E.

$$a \frac{d^2y}{dx^2} + b \frac{dy}{dx} + cy = 0$$

② the general solution of the nonhomogeneous D.E.
is given by

$$y = y_h + y_p$$

where

y_h is the general solution of the homogeneous D.E.

and y_p is a particular solution of the nonhomogeneous D.E.

③ we find y_h by using the methods of
the preceding chapter

④ we find y_p by using a method called
the method of undetermined coefficients

Second-Order Nonhomogeneous Linear D.E.'s

<u>Chpt 19</u> <u>With Constant Coefficients</u>

<u>The Non homogeneous Term</u>

① the method of undetermined coefficients
may be used when the non homogeneous term
$d(x)$ is one of the following types of function

① constant

② polynomial

③ exponential

④ trigonometric

⑤ hyperbolic

<u>constant</u>

$d(x) = K$ where K is a constant

<u>polynomial</u>

$d(x) = K x^n$ $n = 1, 2, 3, \dots$ and K is a constant

<u>exponential</u>

$d(x) = K e^{ax}$ where K and a are constants

<u>trigonometric</u>

$d(x) = K \sin ax$ where K and a are constants

$d(x) = K \cos ax$

<u>hyperbolic</u>

$d(x) = K \sinh ax$ where K and a are constants

$d(x) = K \cosh ax$

Second-Order Nonhomogeneous Linear D.E.'s
Chapt 19 With constant coefficients

① note

all of the preceding nonhomogeneous terms
fall into one of the following two categories

① the function has a finite number of
nonzero derivatives   or

② the derivatives keep repeating

Second-Order Nonhomogeneous Linear D.E.'s
chapt 19 With Constant Coefficients
the function has a finite number of derivatives

ex consider the function $f(x) = 5x^3$

$$f(x) = 5x^3 \quad = 5 \cdot x^3$$
$$f'(x) = 15x^2 \quad = 15 \cdot x^2$$
$$f''(x) = 30x \quad = 30 \cdot x$$
$$f'''(x) = 30 \quad = 30 \cdot 1$$
$$f^{IV}(x) = 0$$
$$f^{V}(x) = 0$$
etc

① the function and its derivatives
form a family which we designate
using set notation

② the family of the function and its derivatives
is given by
$$\{x^3, x^2, x, 1\}$$

Second-Order Nonhomogeneous Linear D.E.'
<u>Chapter 19 With Constant Coefficients</u>
<u>the derivatives keep repeating</u>

ex consider the function $f(x) = 3e^{2x}$

$f(x) = 3e^{2x}$

$f'(x) = 3e^{2x} \cdot 2 = 6e^{2x}$

$f''(x) = 6e^{2x} \cdot 2 = 12e^{2x}$

$f'''(x) = 12e^{2x} \cdot 2 = 24e^{2x}$

$f^{IV}(x) = 24e^{2x} \cdot 2 = 48e^{2x}$

etc

the function and its derivatives
$f = ?$ in the family
$\{e^{2x}\}$

Second-Order Nonhomogeneous Linear D.E. Intro

Chapt 19 With Constant Coefficients

ex consider the function $f(x) = 3 \sin 2x$

$f(x) = 3 \sin 2x$

$f'(x) = 3 \cos 2x \cdot 2 = 6 \cos 2x$

$f''(x) = 6(-\sin 2x)(2) = -12 \sin 2x$

$f'''(x) = -12 \cos 2x (2) = -24 \cos 2x$

$f^{IV}(x) = -24(-\sin 2x)(2) = 48 \sin 2x$

the function and its derivatives

$f^{(n)}$ in the family

$\{\sin 2x, \cos 2x\}$

Second-Order Nonhomogeneous Linear D.E. ✓ Intro.

Chapt 19 With Constant Coefficients

ex consider the function $f(x) = -5 \cos 2x$

$$f(x) = -5 \cos 2x$$
$$f'(x) = -5(-\sin 2x)(2) = 10 \sin 2x$$
$$f''(x) = 10 \cos 2x (2) = 20 \cos 2x$$
$$f'''(x) = 20(-\sin 2x)(2) = -40 \sin 2x$$
$$f^{IV}(x) = -40 \cos 2x (2) = -80 \cos 2x$$

etc

the function and its derivatives

for in the family

$$\{\cos 2x, \sin 2x\}$$

or $\{\sin 2x, \cos 2x\}$

Second-Order Nonhomogeneous Linear D.E.'s With Constant Coefficients

Ex consider the function $f(x) = 2 \sinh 3x$

$f(x) = 2 \sinh 3x$

$f'(x) = 2 \cosh 3x (3) = 6 \cosh 3x$

$f''(x) = 6 \sinh 3x (3) = 18 \sinh 3x$

$f'''(x) = 18 \cosh 3x (3) = 54 \cosh 3x$

$f^{IV}(x) = 54 \sinh 3x (3) = 162 \cosh 3x$

etc

the function and its derivatives

are in the family

$\{ \sinh 3x, \cosh 3x \}$

Second-Order Nonhomogeneous Linear D.E.'s

Chapter 19 With Constant Coefficients

ex Consider the function $6(x) = -3 \cosh x$

$6(x) = -3 \cosh x$

$6'(x) = -3 \sinh x$

$6''(x) = -3 \cosh x$

$6'''(x) = -3 \sinh x$

$6^{IV}(x) = -3 \cosh x$

etc

the function and its derivatives

form the family

$[\cosh x, \sinh x]$

or $[\sinh x, \cosh x]$

Second-Order Nonhomogeneous Linear D.E's

Chapt 19 With Constant Coefficient

③ we may write the following functions
and their families

① $f(x) = K$ function
     $\{1\}$ family

② $f(x) = Kx$ function
     $\{x, 1\}$ family

③ $f(x) = Kx^2$ function
     $\{x^2, x, 1\}$ family

④ $f(x) = Kx^3$ function
     $\{x^3, x^2, x, 1\}$ family

⑤ $f(x) = Kx^n$ function
     $\{x^n, x^{n-1}, x^{n-2}, \ldots x^2, x, 1\}$ family

⑥ $f(x) = K e^{ax}$ function
     $\{e^{ax}\}$ family

⑦ $f(x) = K \sin ax$ function
     $\{\sin ax, \cos ax\}$ family

⑧ $f(x) = K \cos ax$ function
     $\{\sin ax, \cos ax\}$ family

intro

Second-Order Nonhomogeneous Linear D.E.'s
chapt 19 With Constant Coefficients

⑨ $f(x) = K \sinh ax$

{ sinhax, coshax }

function family

⑩ $f(x) = K \cosh ax$

{ sinhax, coshax }

function family

note

at this point it is convenient to make a table which includes this information

Second-Order Nonhomogeneous Linear D.E's

Chapt 19 With Constant Coefficients

Making A Table Of The Function

And Its Family

the function	its family
K	$\{1\}$
Kx	$\{x, 1\}$
Kx^2	$\{x^2, x, 1\}$
Kx^3	$\{x^3, x^2, x, 1\}$
Kx^n	$\{x^n, x^{n-1}, x^{n-2}, \dots x^2, x, 1\}$
Ke^{ax}	$\{e^{ax}\}$
$K \sin ax$	$\{\sin ax, \cos ax\}$
$K \cos ax$	$\{\sin ax, \cos ax\}$
$K \sinh ax$	$\{\sinh ax, \cosh ax\}$
$K \cosh ax$	$\{\sinh ax, \cosh ax\}$

Second-Order Non-homogeneous Linear D.E.'s
With Constant Coefficients
The Method of Undetermined Coefficients

① the method of undetermined coefficients involves the following steps —

② finding the family set of the nonhomogeneous term

③ forming a general linear combination of the family members which is set equal to y_p

y_p = the general linear combination of the family members

④ substituting y_p into the nonhomogeneous D.E. to determine the value of the coefficients

④ thus we obtain the particular solution y_p

⑤ since we have already solved the corresponding homogeneous D.E. for y_h we know the general solution of the nonhomogeneous D.E. given by

$y = y_h + y_p$

⑥ the following examples illustrate this technique

(377)

Second - Order Nonhomogeneous Linear D.E.'s
Chapt 19 With Constant Coefficients

① solve the D.E.

$$\frac{d^2y}{dx^2} - 5\frac{dy}{dx} + 6y = 7$$

$$\frac{d^2y}{dx^2} - 5\frac{dy}{dx} + 6y = 0$$

$$m^2 - 5m + 6 = 0$$

$$(m-3)(m-2) = 0$$

$$m-3 = 0 \qquad m-2 = 0$$

$$m = 3 \qquad m = 2$$

$$m = 2, 3$$

$$y = C_1 e^{m_1 x} + C_2 e^{m_2 x}$$

$$y = C_1 e^{2x} + C_2 e^{3x}$$

$$y_h = C_1 e^{2x} + C_2 e^{3x}$$

$$y = y_h + y_p$$

Second-Order Nonhomogeneous Linear D.E.'s
chapt 19 with constant Coefficients

(cont) finding y_p

the nonhomogeneous term is $g(x) = 7$

its family is

$$\{1\}$$

$y_p = A \cdot 1$

$y_p = A$

$y'_p = 0$

$y''_p = 0$

$y'' - 5y' + 6y = 7$

$0 - 0 + 6A = 7$

$6A = 7$

$A = \dfrac{7}{6}$

$y_p = A$

$y_p = \dfrac{7}{6}$

$y = y_h + y_p$

$y = c_1 e^{2x} + c_2 e^{3x} + \dfrac{7}{6}$

I C

Second-Order Nonhomogeneous Linear D.E.'s

Chapt 19 With Constant Coefficients

(1 cont) check $y_p = \frac{7}{6}$

$$y'' - 5y' + 6y = 7$$

$$y_p = \frac{7}{6}$$
$$y'_p = 0$$
$$y''_p = 0$$

$$0 - 0 + 6\left(\frac{7}{6}\right) = 7$$
$$7 = 7$$

Second-Order Nonhomogeneous Linear D.E.'s
Chapt 19 With Constant Coefficients

(E) solve the D.E.

$$\frac{d^2 y}{dx^2} - 4y = 5x$$

$$\frac{d^2 y}{dx^2} - 4y = 0$$

$$m^2 - 4 = 0$$

$$m^2 = 4$$

$$m = \pm \sqrt{4}$$

$$m = \pm 2$$

$$m = 2, -2$$

$$y = C_1 e^{m_1 x} + C_2 e^{m_2 x}$$

$$y = C_1 e^{2x} + C_2 e^{-2x}$$

$$y_h = C_1 e^{2x} + C_2 e^{-2x}$$

$$y = y_h + y_p$$

Second- Order Nonhomogeneous Linear D.E.'s
with Constant Coefficients

(2 cont) finding y_p

the nonhomogeneous term $\sim 5x$

its family $\sim$

$$\{x, 1\}$$

$$y_p = Ax + B \cdot 1$$
$$y_p = Ax + B$$
$$y'_p = A$$
$$y''_p = 0$$

$$y'' - 4y = 5x$$
$$0 - 4(Ax + B) = 5x$$
$$-4Ax - 4B = 5x$$
$$-4A = 5 \qquad -4B = 0$$
$$A = -\frac{5}{4} \qquad B = 0$$

$$y_p = Ax + B$$
$$y_p = -\frac{5}{4}x + 0$$
$$y_p = -\frac{5}{4}x$$

Second-Order Nonhomogeneous Linear D.E.:
Chapt 19   With Constant Coefficients

(2 cont)

$$y = y_h + y_p$$

$$y = c_1 e^{2x} + c_2 e^{-2x} - \frac{5}{4} x$$

Second Order Nonhomogeneous Linear D.E.'s
Chapt 19 With Constant Coefficient

(2 cont) check $y_p = -\frac{5}{4}x$

$$y'' - 4y = 5x$$

$$y_p = -\frac{5}{4}x$$

$$y'_p = -\frac{5}{4}$$

$$y''_p = 0$$

$$0 - 4\left(-\frac{5}{4}x\right) = 5x$$

$$5x = 5x$$

III A

Second-Order Nonhomogeneous Linear D.E.;
Chapt 19 With Constant Coefficients

(3) solve the D.E.

$$\frac{d^2 y}{dx^2} + 3 \frac{dy}{dx} + 2y = -3x^2$$

$$\frac{d^2 y}{dx^2} + 3 \frac{dy}{dx} + 2y = 0$$

$$m^2 + 3m + 2 = 0$$

$$(m+1)(m+2) = 0$$

$$m+1 = 0 \qquad m+2 = 0$$

$$m = -1 \qquad m = -2$$

$$m = -1, -2$$

$$y = c_1 e^{m_1 x} + c_2 e^{m_2 x}$$

$$y = c_1 e^{-x} + c_2 e^{-2x}$$

$$y_h = c_1 e^{-x} + c_2 e^{-2x}$$

$$y = y_h + y_p$$

Second-Order Nonhomogeneous Linear D.E.'s
Chapter 19 With Constant Coefficients

(3cont) finding y_p

the nonhomogeneous term is $-3x^2$

its family is
$$\{x^2, x, 1\}$$

$$y_p = Ax^2 + Bx + C \cdot 1$$
$$y_p = Ax^2 + Bx + C$$
$$y'_p = 2Ax + B$$
$$y''_p = 2A$$

$$y'' + 3y' + 2y = -3x^2$$
$$2A + 3(2Ax + B) + 2(Ax^2 + Bx + C) = -3x^2$$
$$2A + 6Ax + 3B + 2Ax^2 + 2Bx + 2C = -3x^2$$
$$2Ax^2 + 6Ax + 2Bx + 2A + 3B + 2C = -3x^2$$
$$2Ax^2 + (6A + 2B)x + (2A + 3B + 2C) = -3x^2$$

$$2A = -3 \qquad 6A + 2B = 0 \qquad 2A + 3B + 2C = 0$$

$$A = -\frac{3}{2} \qquad 6\left(-\frac{3}{2}\right) + 2B = 0 \qquad 2\left(-\frac{3}{2}\right) + 3\left(\frac{9}{2}\right) + 2C = 0$$

$$-9 + 2B = 0 \qquad -\frac{6}{2} + \frac{27}{2} + 2C = 0$$

$$2B = 9 \qquad \frac{21}{2} + 2C = 0$$

$$B = \frac{9}{2} \qquad 2C = \frac{21}{2}$$

$$C = \frac{21}{4}$$

$$A = -\frac{3}{2}, \quad B = \frac{9}{2}, \quad C = \frac{21}{4}$$

Second- Order Nonhomogeneous Linear D.E.'s

Chapter 19 With Constant Coefficients

3 cont $y_p = A x^2 + B x + C$

$y_p = -\frac{3}{2} x^2 + \frac{9}{2} x + \frac{21}{4}$

$y = y_h + y_p$

$y = C_1 e^{-x} + C_2 e^{-2x} - \frac{3}{2} x^2 + \frac{9}{2} x + \frac{21}{4}$

III V

Second- Order Nonhomogeneous Linear D.E.'s

Chapt 19 With Constant Coefficients

(3 cont)

Check $y_p = -\frac{3}{2}x^2 + \frac{9}{2}x + \frac{21}{4}$

$$y'' + 3y' + 2y = -3x^2$$

$$y_p = -\frac{3}{2}x^2 + \frac{9}{2}x + \frac{21}{4}$$

$$y'_p = -3x + \frac{9}{2}$$

$$y''_p = -3$$

$$-3 + 3\left(-3x + \frac{9}{2}\right) + 2\left(-\frac{3}{2}x^2 + \frac{9}{2}x + \frac{21}{4}\right) = -3x^2$$

$$3 - 9x + \frac{27}{2} - 3x^2 + 9x + \frac{21}{2} = -3x^2$$

$$\frac{6}{2} - 9x + \frac{27}{2} - 3x^2 + 9x + \frac{21}{2} = -3x^2$$

$$-\frac{\cancel{27}}{2} - \cancel{9x} - 3x^2 + \cancel{9x} + \frac{\cancel{21}}{2} = -3x^2$$

$$-3x^2 = -3x^2$$

Second-Order Nonhomogeneous Linear DEs
chapt 19 With constant coefficients

① solve the first-order DE

$$\frac{dy}{dx} + y = x$$

The method using two different methods

of ① the general 1st-order linear D.E method and
Undetermine ② the method of undetermined coefficients
coefficients

1st order method ← the general 1st-order linear D.E method
DE ↓ $\quad \frac{dy}{dx} + y = x$

$$\left[\begin{array}{l} \frac{dy}{dx} + P(x)\,y = Q(x) \\[2mm] M = e^{\int P\,dx} \\[2mm] My = \int M Q\, dx \end{array}\right.$$

$$\frac{dy}{dx} + y = x$$

$$\frac{dy}{dx} + P(x)\,y = Q(x)$$

$$\quad\quad P(x) = 1 \quad\quad Q(x) = x$$

$$M = e^{\int P\,dx}$$
$$M = e^{\int 1\,dx}$$
$$M = e^{x}$$

I B   Prob

<u>Second-Order Nonhomogeneous Linear D.E.</u>

Chapt 19 With constant Coefficients

(Cont) $\mu y = \int \mu Q\, dx$

$e^x y = \int e^x \cdot x \; dx$

The $e^x y = \int x e^x\, dx$

Method

Of $\int x e^x\, dx$

Undetermined

Coeficients $\int u\, dv = uv - \int v\, du$

1st   let $u = x$ $dv = e^x\, dx$

Order $du = dx$ $\int dv = \int e^x\, dx$

D.E. $v = e^x$

$\int x e^x\, dx$

$= x e^x - \int e^x\, dx$

$= x e^x - (e^x + c)$

$= x e^x - e^x + C$

$e^x y = x \cdot e^x - e^x + C$

$\dfrac{e^x y}{e^x} = \dfrac{x e^x - e^x + C}{e^x}$

$y = \dfrac{x e^x}{e^x} - \dfrac{e^x}{e^x} + \dfrac{C}{e^x}$

$y = x - 1 + C e^{-x}$

$y = C e^{-x} + x - 1$

Second Order Nonhomogeneous Linear D.E's

Chpt 19 With Constant Coefficients

(1 cont) Check

$$\frac{dy}{dx} + y = x$$

The Method Of Undetermined Coefficients 1st Order DEs

$$y = Ce^{-x} + x - 1$$

$$\frac{dy}{dx} = Ce^{-x}(-1) + 1 - 0$$

$$\frac{dy}{dx} = -Ce^{-x} + 1$$

$$(-Ce^{-x} + 1) + (Ce^{-x} + x - 1) = x$$

$$-Ce^{-x} + 1 + Ce^{-x} + x - 1 = x$$

$$x = x$$

Second-Order Nonhomogeneous Linear DE's

Chapter 19 With Constant Coefficients

(cont) method 2: the method of undetermined coefficient

$$\frac{dy}{dx} + y = x$$

The

Method

of

Undetermined

Coefficient

Let

Order

DE's

$$\frac{dy}{dx} + y = 0$$

let $\quad y = Ce^{mx}$

$\quad\quad y' = Ce^{mx} \cdot m$

$\quad\quad y' = mCe^{mx}$

$$mCe^{mx} + Ce^{mx} = 0$$

$$Ce^{mx}(m+1) = 0$$

$$m + 1 = 0$$

$$m = -1$$

$$y = Ce^{mx}$$
$$y = Ce^{(-1)x}$$
$$y = Ce^{-x}$$

also

$$\frac{dy}{dx} + y = 0 \qquad\qquad y = Ce^{mx}$$
$$\qquad\qquad\qquad\qquad y = Ce^{(-1)x}$$
$$m + 1 = 0 \qquad\qquad y = Ce^{-x}$$
$$m = -1$$

$$y = y_h + y_p$$

Second-Order Nonhomogeneous Linear D.E.
Ch. 19 With Constant Coefficients

(cont) Finding y_p

the family of the nonhomogeneous term
x is given by

The
Method
of {x, 1}

Undetermined $y_p = Ax + B \cdot 1$
Coefficients $y_p = Ax + B$ ✓
1st $y_p' = A$ ✓

Order
D.E. $\frac{dy}{dx} + y = x$

$A + (Ax + B) = x$

$A + Ax + B = x$

$Ax + A + B = x$

$Ax + (A + B) = x$

$A = 1 \qquad A + B = 0$

$\qquad\qquad 1 + B = 0$

$\qquad\qquad B = -1$

$A = 1, \quad B = -1$

$y_p = Ax + B$

$y_p = (1)x + (-1) = x - 1$

$y = y_h + y_p$

$y = Ce^{-x} + x - 1$

Second-Order Nonhomogeneous Linear D.E.
Chapt 19 With Constant Coeficients

note
we do not need the method of undetermined coeficients
to solve 1st-order nonhomogeneous linear D.E.
with constant coeficients

The
Method
of
Undetermined   however
Coeficents   we do need the method of undetermined coeficients
1st   to solve 2nd-order nonhomogeneous linear D.E
Order   with constant coeficients
D.E.

1st - order   do not need M.O.U.C.
$$a \frac{dy}{dx} + by = c(x)$$

2nd - order   do need M.O.U.C.
$$a \frac{d^2y}{dx^2} + b \frac{dy}{dx} + cy = d(x)$$

note also
$c(x)$ and $d(x)$ may be functions that are
constants or just plain constants

Second-Order Nonhomogeneous Linear D.E.'s
Chpt 19 With Constant Coefficients

① consider the D.E

$$a \frac{d^2 y}{dx^2} + b \frac{dy}{dx} + cy = d(x)$$

prove

if

y_h is a solution to the homogeneous D.E.

$$a \frac{d^2 y}{dx^2} + b \frac{dy}{dx} + cy = 0$$

and

y_p is a particular solution to the nonhomogeneous D.E.

$$a \frac{d^2 y}{dx^2} + b \frac{dy}{dx} + cy = d(x)$$

then

$y = y_h + y_p$ is a solution to

$$a \frac{d^2 y}{dx^2} + b \frac{dy}{dx} + cy = d(x)$$

Second-Order Nonhomogeneous Linear D.E.:

Chapt 19 with Constant Coefficients

(cont) y_h is a solution to the homogeneous D.E.

$$a \frac{d^2 y}{dx^2} + b \frac{dy}{dx} + c y = 0$$

proof ① $a \frac{d^2 y_h}{dx^2} + b \frac{dy_h}{dx} + c y_h = 0$

y_p is a solution to the nonhomogeneous D.E.

$$a \frac{d^2 y}{dx^2} + b \frac{dy}{dx} + c y = d(x)$$

② $a \frac{d^2 y_p}{dx^2} + b \frac{dy_p}{dx} + c y_p = d(x)$

add equations ① and ②

Second-Order Nonhomogeneous Linear D.E.'s

Chapt 19 With Constant Coefficients

$\boxed{1\,cont}$ $a\dfrac{d^2 y_h}{dx^2} + b\dfrac{dy_h}{dx} + c\,y_h = 0$

$a\dfrac{d^2 y_p}{dx^2} + b\dfrac{dy_p}{dx} + c\,y_p = d(x)$

proof

$a\dfrac{d^2 y_h}{dx^2} + a\dfrac{d^2 y_p}{dx^2}$

$+\ b\dfrac{dy_h}{dx} + b\dfrac{dy_p}{dx}$

$+\ c\,y_h + c\,y_p$

$= d(x)$

$a\left(\dfrac{d^2 y_h}{dx^2} + \dfrac{d^2 y_p}{dx^2}\right)$

$+\ b\left(\dfrac{dy_h}{dx} + \dfrac{dy_p}{dx}\right)$

$+\ c\,(y_h + y_p)$

$= d(x)$

$a\dfrac{d^2}{dx^2}(y_h + y_p)$

$+\ b\dfrac{d}{dx}(y_h + y_p)$

$+\ c\,(y_h + y_p)$

$= d(x)$

I V

Prob

Second - Order Nonhomogeneous Linear D.E.'s

chapt 19   With Constant Coefficients

(cont)

$$a \frac{d^2 y}{dx^2} + b \frac{dy}{dx} + c y = d(x)$$

where

$$y = y_n + y_p$$

Proof

<u>Unit 20 The Nonhomogeneous Term As A Sum Or A Product</u>

<u>Two Special Cases For The Nonhomogeneous Term</u>

① so far we have considered the nonhomogeneous term

of the D.E

$$a \frac{d^2 y}{dx^2} + b \frac{dy}{dx} + c y = d(x)$$

to be a single function

② however we will consider the following two

special cases -

ⓐ the nonhomogeneous term is a sum (or difference)

such as $3x + 7e^x$ or $3x^2 - 7x$

ⓑ the nonhomogeneous term is a product

such as $5 x e^x$

<u>the sum (or difference) of two functions</u>

to find its family take the union of the sets

of each function

<u>the product of two functions</u>

to find its family multiply the members of one set

by the members of the other set

Cont'd The Nonhomogeneous Term Is A Sum Or A Product

sums and differences

ex) find the family of $f(x) = 5x^2 + 3e^{5x}$

the family of $5x^2$ is
$\{x^2, x, 1\}$

the family of $3e^{5x}$ is
$\{e^{5x}\}$

the family of $5x^2 + 3e^{5x}$ is
$\{x^2, x, 1\} \cup \{e^{5x}\}$
$= \{x^2, x, 1, e^{5x}\}$

ex) find the family of $f(x) = 15x^2 - 2x$

the family of $15x^2$ is
$\{x^2, x, 1\}$

the family of $-2x$ is
$\{x, 1\}$

the family of $15x^2 - 2x$ is
$\{x^2, x, 1\} \cup \{x, 1\}$
$= \{x^2, x, 1\}$

Chapt 20 The Nonhomogeneous Term f A Sum Or A Product

ex find the family of $f(x) = 3 \sin 2x - 4 \cos 2x$

the family of $3 \sin 2x$ is
$\{\sin 2x, \cos 2x\}$

the family of $-4 \cos 2x$ is
$\{\cos 2x, \sin 2x\}$
or $\{\sin 2x, \cos 2x\}$

the family of $3 \sin 2x + (-4 \cos 2x)$ is
$\{\sin 2x, \cos 2x\} \cup \{\sin 2x, \cos 2x\}$
$= \{\sin 2x, \cos 2x\}$

Chapt 20 The Nonhomogeneous Term Is A Sum Or A Product

product

1x find the family of $f(x) = 3x^2 e^{-3x}$

the family of $3x^2$ is

$\{x^2, x, 1\}$

the family of e^{-3x} is

$\{e^{-3x}\}$

the family of $3x^2 e^{-3x}$ is

$\{x^2, x, 1\} \times \{e^{-3x}\}$

$= \{x^2 e^{-3x}, x e^{-3x}, e^{-3x}\}$

2x find the family of $f(x) = 6 e^{2x} \sin 3x$

the family of $6 e^{2x}$ is

$\{e^{2x}\}$

the family of $\sin 3x$ is

$\{\sin 3x, \cos 3x\}$

the family of $6 e^{2x} \sin 3x$ is

$\{e^{2x}\} \times \{\sin 3x, \cos 3x\}$

$= \{e^{2x} \sin 3x, e^{2x} \cos 3x\}$

$\overline{V}$

(402)

Chapter 20 The Non Homogeneous Term Is A Sum Or A Product

the non homogeneous term is a sum

ex   solve the   D.E.

$$\frac{d^2y}{dx^2} + \frac{dy}{dx} - 2y = x + 3$$

$$\frac{d^2y}{dx^2} + \frac{dy}{dx} - 2y = 0$$

$$m^2 + m - 2 = 0$$

$$(m+2)(m-1) = 0$$

$$m+2 = 0 \qquad m-1 = 0$$

$$m = -2 \qquad m = 1$$

$$m = 1, -2$$

$$y = C_1 e^{m_1 x} + C_2 e^{m_2 x}$$

$$y = C_1 e^x + C_2 e^{-2x}$$

$$y_h = C_1 e^x + C_2 e^{-2x}$$

$$y = y_h + y_p$$

(403)

Chapt 20 The Non homogeneous Term is A Sum Or A Product

~~finding~~ y_p

the non homogeneous term is $b(x) = x + 3$

the family of x is

$\{x, 1\}$

the family of 3 is

$\{1\}$

the family of $x + 1$ is

$\{x, 1\} \cup \{1\}$

$= \{x, 1\}$

$y_p = A x + B \cdot 1$

$y_p = A x + B$

$y'_p = A$

$y''_p = 0$

$y'' + y' - 2y = x + 3$

$0 + A - 2(A x + B) = x + 3$

$A - 2 A x - 2B = x + 3$

$-2 A x + A - 2B = x + 3$

$-2 A x + (A - 2B) = x + 3$

$-2 A = 1 \qquad\qquad A - 2B = 3$

$\quad A = -\dfrac{1}{2} \qquad -\dfrac{1}{2} - 2B = 3$

$\qquad\qquad\qquad\qquad 2B = -\dfrac{1}{2} - 3$

$\qquad\qquad\qquad\qquad 2B = -\dfrac{1}{2} - \dfrac{6}{2}$

$\qquad\qquad\qquad\qquad 2B = -\dfrac{7}{2}$

$\qquad\qquad\qquad\qquad B = -\dfrac{7}{4}$

(404)

Chapt 20 The Nonhomogeneous Term Is A Sum Or A Product

$y_r = Ax + B$

$A = -\frac{1}{2} \qquad B = -\frac{7}{4}$

$y_p = -\frac{1}{2}x - \frac{7}{4}$

$y = y_h + y_p$

$y = c_1 e^x + c_2 e^{-2x} - \frac{1}{2}x - \frac{7}{4}$

(405)

Convert The Nonhomogeneous Term to A Sum Or A Product

Check $y_p = -\frac{1}{2}x - \frac{1}{4}$

$$y'' + y' - 2y = x + 3$$

$$y_p = -\frac{1}{2}x - \frac{1}{4}$$
$$y'_p = -\frac{1}{2}$$
$$y''_p = 0$$

$$0 - \frac{1}{2} - 2\left(-\frac{1}{2}x - \frac{1}{4}\right) = x + 3$$

$$-\frac{1}{2} + x + \frac{1}{2} = x + 3$$

$$x + \frac{6}{2} = x + 3$$

$$x + 3 = x + 3$$

(406)

Chapter 20 The Nonhomogeneous Term Is A Sum Or A Product
the nonhomogeneous term is a product
&x solve the D.E.

$$\frac{d^2y}{dx^2} - 4y = xe^x$$

$$\frac{d^2y}{dx^2} - 4y = 0$$

$$m^2 - 4 = 0$$
$$m^2 = 4$$
$$m = \pm \sqrt{4}$$
$$m = \pm 2$$

$$y = C_1 e^{m_1 x} + C_2 e^{m_2 x}$$
$$y = C_1 e^{2x} + C_2 e^{-2x}$$

$$y = y_h + y_p$$

Ch.20 The Nonhomogeneous Term is A Sum Or A Product

finding y_p

the non homogeneous term is $x e^x$

the family of x is

$\{x, 1\}$

the family of e^x is

$\{e^x\}$

the family of $x e^x$ is

$\{x, 1\} \times \{e^x\}$

$= \{x e^x, e^x\}$

$y_p = A x e^x + B e^x$

$y'_p = A[x e^x + e^x(1)] + B e^x$

$y'_p = A x e^x + A e^x + B e^x$

$y''_p = A[x e^x + e^x(1)] + A e^x + B e^x$

$y''_p = A x e^x + A e^x + A e^x + B e^x$

$y''_p = A x e^x + 2 A e^x + B e^x$

$y''_p = A x e^x + (2A + B) e^x$

(408)

Chapter 20 The Nonhomogeneous Term Is A Sum Or A Product

$$y'' - 4y = xe^x$$

$$Axe^x + (2A+B)e^x$$
$$-4(Axe^x + Be^x)$$
$$= xe^x$$

$$Axe^x + (2A+B)e^x$$
$$+ 4Axe^x - 4Be^x$$
$$= xe^x$$

$$Axe^x - 4Axe^x + 2Ae^x + Be^x - 4Be^x$$
$$= xe^x$$

$$-3Axe^x + (2A-3B)e^x = xe^x$$

$$-3A = 1 \qquad 2A - 3B = 0$$
$$A = -\frac{1}{3} \qquad 2\left(-\frac{1}{3}\right) - 3B = 0$$
$$-\frac{2}{3} - 3B = 0$$
$$3B = -\frac{2}{3}$$
$$B = -\frac{2}{9}$$

$$A = -\frac{1}{3}, \quad B = -\frac{2}{9}$$

$$y_p = Axe^x + Be^x$$
$$y_p = -\frac{1}{3}xe^x - \frac{2}{9}e^x$$

(409)

Chapt 20 The Nonhomogeneous Term for A Sum On A Product

$$y = y_h + y_p$$

$$y = c_1 e^{2x} + c_2 e^{-2x} - \frac{1}{3} x e^x - \frac{2}{9} e^x$$

(410)

Chapt 20 The Nonhomogeneous Term $y =$ A Sum Or A Product

check $\quad y_p = -\frac{1}{3}x e^x - \frac{2}{9}e^x$

$y'' - 4y = x e^x$

$y_p = -\frac{1}{3}x e^x - \frac{2}{9}e^x$

$y'_p = -\frac{1}{3}[x e^x + e^x(1)] - \frac{2}{9}e^x$

$y'_p = -\frac{1}{3}x e^x - \frac{1}{3}e^x - \frac{2}{9}e^x$

$y'_p = -\frac{1}{3}x e^x - \frac{3}{9}e^x - \frac{2}{9}e^x$

$y'_p = -\frac{1}{3}x e^x - \frac{5}{9}e^x$

$y''_p = -\frac{1}{3}[x e^x + e^x(1)] - \frac{5}{9}e^x$

$y''_p = -\frac{1}{3}x e^x - \frac{1}{3}e^x - \frac{5}{9}e^x$

$y''_p = -\frac{1}{3}x e^x - \frac{3}{9}e^x - \frac{5}{9}e^x$

$y''_p = -\frac{1}{3}x e^x - \frac{8}{9}e^x$

$-\frac{1}{3}x e^x - \frac{8}{9}e^x - 4\left(-\frac{1}{3}x e^x - \frac{2}{9}e^x\right)$

$= x e^x$

$-\frac{1}{3}x e^x - \cancel{\frac{8}{9}e^x} + \frac{4}{3}x e^x + \cancel{\frac{8}{9}e^x}$

$= x e^x$

$\frac{3}{3}x e^x = x e^x$

$x e^x = x e^x$

(411)

Chapter 20 The Nonhomogeneous Term Is A Sum Or A Product

one Final Rule

① the method of undetermined coefficients has
one final rule -
no family member of the non homogeneous term
may be included in the general solution of
the corresponding homogeneous D. E.

② If a family member of the non homogeneous term
is included in the general solution of
the homogeneous D. E. multiply each family member
by x

③ If a family member is still included in
the general solution of the homogeneous D.E
multiply each family member again by x

④ this procedure may be repeated until
no family member of the non homogeneous term
is included in the general solution of the
homogeneous D.E.

Chpt 2D The Nonhomogeneous Term y_0 A Sum Or A Product

ex solve the D. E.

$$\frac{d^2 y}{dx^2} + y = \sin x$$

$$\frac{d^2 y}{dx^2} + y = 0$$

$$m^2 + 1 = 0$$

$$m^2 = -1$$

$$m = \pm \sqrt{-1}$$

$$m = \pm i$$

$$0 \pm i$$

$$a \pm b i$$

$$a = 0 \qquad b = 1$$

$$y = e^{ax}(C_1 \cos bx + C_2 \sin bx)$$
$$y = e^0 (C_1 \cos x + C_2 \sin x)$$
$$y = 1 (C_1 \cos x + C_2 \sin x)$$
$$y = C_1 \cos x + C_2 \sin x$$

$$y = y_h + y_p$$

Chapt 20 The Nonhomogeneous Term to A Sum Or A Product finding y_P

the nonhomogeneous term $\to$ $g(x) = \sin x$

to family is

$\{\sin x, \cos x\}$

however

$\sin x$ and $\cos x$ are both included in the general solution of the corresponding homogeneous D.E.

therefore multiply each family member by x

$\{x \sin x, x \cos x\}$

$y_P = A x \sin x + B x \cos x$

$y'_P = A \left[x \cos x + \sin x (1) \right] + B \left[x (-\sin x) + \cos x (1) \right]$

$y'_P = A x \cos x + A \sin x - B x \sin x + B \cos x$

$y'_P = A x \cos x - B x \sin x + A \sin x + B \cos x$

$y''_P = A \left[x (-\sin x) + \cos x (1) \right] - B \left[x \cos x + \sin x (1) \right]$
$\qquad + A \cos x + B (-\sin x)$

$y''_P = - A x \sin x + A \cos x - B x \cos x - B \sin x$
$\qquad + A \cos x - B \sin x$

$y''_P = - A x \sin x - B x \cos x + 2A \cos x - 2B \sin x$

(414)

Chapt 20 The Non homogeneous Term ↓ A Sum Or A Product

$$y'' + y = \sin x$$

$$- Ax \sin x - Bx \cos x + 2A \cos x - 2B \sin x$$
$$+ Ax \sin x + Bx \cos x$$
$$= \sin x$$

$$2A \cos x - 2B \sin x = \sin x$$
$$2A = 0 \qquad -2B = 1$$
$$A = 0 \qquad B = -\frac{1}{2}$$

$$y_p = Ax \sin x + Bx \cos x$$
$$y_p = 0 - \frac{1}{2} x \cos x$$
$$y_p = -\frac{1}{2} x \cos x$$

$$y = y_h + y_p$$
$$y = c_1 \cos x + c_2 \sin x - \frac{1}{2} x \cos x$$

415

Chapt 20 The Nonhomogeneous Term Yo A Sum On A Product

check $y_p = -\frac{1}{2} x \cos x$

$$y'' + y = \sin x$$

$$y_p = -\frac{1}{2} x \cos x$$

$$y'_p = -\frac{1}{2}\left[x(-\sin x) + \cos x (1)\right]$$

$$y'_p = \frac{1}{2} x \sin x - \frac{1}{2} \cos x$$

$$y''_p = \frac{1}{2}\left[x \cos x + \sin x (1)\right] - \frac{1}{2}(-\sin x)$$

$$y''_p = \frac{1}{2} x \cos x + \frac{1}{2} \sin x + \frac{1}{2} \sin x$$

$$y''_p = \frac{1}{2} x \cos x + \sin x$$

$$\left(\frac{1}{2} x \cos x + \sin x\right) + \left(-\frac{1}{2} x \cos x\right) = \sin x$$

$$\frac{1}{2} x \cos x + \sin x - \frac{1}{2} x \cos x = \sin x$$

$$\sin x = \sin x$$

(416)

Chpt 21 Second-Order Nonhomogeneous Initial value Problems
Solving Second - Order Nonhomogeneous
Initial Value Problems (I. V. P's)

① now that we are able to solve
second-order nonhomogeneous linear D.E.'s
with constant coefficients using the method
of undetermined coefficients we will
turn our attention to solving initial value problems

② two methods for solving I. V. P's will be used

Ⓐ the conventional method

Ⓑ using Laplace transforms

$$\boxed{417}$$

Chapt 21 Second-Order Nonhomogeneous Initial Value Problems

conventional method

solve the initial value problem

$$\frac{d^2 y}{dx^2} - y = x$$

$$y(0) = 1$$

$$y'(0) = -3$$

$$\frac{d^2 y}{dx^2} - y = 0$$

$$m^2 - 1 = 0$$

$$m^2 = 1$$

$$m = \pm \sqrt{1}$$

$$m = \pm 1$$

$$m = 1, -1$$

$$y = c_1 e^{m_1 x} + c_2 e^{m_2 x}$$

$$y = c_1 e^{x} + c_2 e^{-x}$$

$$y_h = c_1 e^{x} + c_2 e^{-x}$$

$$y = y_h + y_p$$

(418)

Chpt 21 Second-Order Nonhomogeneous Initial Value Problems

finding y_p

the nonhomogeneous term is x

its family is

$$[x, 1]$$

$$y_p = Ax + B \cdot 1$$
$$y_p = Ax + B$$
$$y'_p = A$$
$$y''_p = 0$$

$$y'' - y = x$$
$$0 - (Ax + B) = x$$
$$-Ax - B = x$$
$$-A = 1 \qquad -B = 0$$
$$A = -1 \qquad B = 0$$

$$y_p = Ax + B$$
$$y_p = (-1)x$$
$$y_p = -x$$

Chapt 21 Second-Order Nonhomogeneous Initial Value Problem

$$y = y_h + y_p$$
$$y = c_1 e^x + c_2 e^{-x} - x$$

(420)

Chapter 21 Second-Order Nonhomogeneous Initial Value Problems

check $y_p = -x$

$$y'' - y = x$$

$$y_p = -x$$
$$y'_p = -1$$
$$y''_p = 0$$

$$0 - (-x) = x$$
$$0 + x = x$$
$$x = x$$

(421)

Chapter 21 Second-Order Nonhomogeneous Initial Value Problems

$\underline{y(0) = 1}$ $x = 0, \quad y = 1$

$y = c_1 e^x + c_2 e^{-x} - x$

$1 = c_1 e^0 + c_2 e^0 - 0$

$1 = c_1 (1) + c_2 (1)$

$1 = c_1 + c_2$

$c_1 + c_2 = 1$

$\underline{y'(0) = -3}$ $x = 0, \quad y' = -3$

$y = c_1 e^x + c_2 e^{-x} - x$

$y' = c_1 e^x - c_2 e^{-x} - 1$

$-3 = c_1 e^0 - c_2 e^0 - 1$

$-3 = c_1 (1) - c_2 (1) - 1$

$-3 = c_1 - c_2 - 1$

$c_1 - c_2 = -3 + 1$

$c_1 - c_2 = -2$

$c_1 + c_2 = 1$

$c_1 - c_2 = -2$

(422)

chapter 21 Second-Order Nonhomogeneous Initial Value Problems

$C_1 + C_2 = 1$

$C_1 - C_2 = -2$

$$C_1 = \frac{\begin{vmatrix} 1 & 1 \\ -2 & -1 \end{vmatrix}}{\begin{vmatrix} 1 & 1 \\ 1 & -1 \end{vmatrix}} = \frac{-1+2}{-1-1} = \frac{1}{-2} = -\frac{1}{2}$$

$$C_2 = \frac{\begin{vmatrix} 1 & 1 \\ 1 & -2 \end{vmatrix}}{\begin{vmatrix} 1 & 1 \\ 1 & -1 \end{vmatrix}} = \frac{-2-1}{-1-1} = \frac{-3}{-2} = \frac{3}{2}$$

$C_1 = -\frac{1}{2}, \quad C_2 = \frac{3}{2}$

$y = C_1 e^x + C_2 e^{-x} - x$

$y = -\frac{1}{2} e^x + \frac{3}{2} e^{-x} - x$

$$\boxed{423}$$

Chapt 21 Second-Order Nonhomogeneous Initial Value Problems using the Laplace Transform

Ex use the Laplace transform to solve the initial value problem $(I.V.P.)$

$$\frac{d^2y}{dx^2} - y = x$$

$$y(0) = 1$$

$$y'(0) = -3$$

$$\frac{d^2y}{dx^2} - y = x$$

$$L\left[\frac{d^2y}{dx^2} - y\right] = L[x]$$

$$L\left[\frac{d^2y}{dx^2}\right] - L[y] = L[x]$$

$$L[y'] = p\,L[y] - y(0)$$

$$L[y''] = p^2\,L[y] - p\,y(0) - y'(0)$$

$$p^2 L[y] - p\,y(0) - y'(0) - L[y] = L[x]$$

$$p^2 L[y] - p\,(1) - (-3) - L[y] = \frac{1!}{p^{1+1}}$$

$$p^2 L[y] - p + 3 - L[y] = \frac{1}{p^2}$$

$$p^2 L[y] - L[y] = \frac{1}{p^2} + p - 3$$

$$L[y]\,(p^2 - 1) = \left(\frac{1}{p^2} + p - 3\right)$$

$$\overline{424}$$

Chapter 21 Second-Order Nonhomogeneous Initial value Problems

$$L[y] = \left(\frac{1}{p^2-1}\right)\left(-\frac{1}{p^2} + p - 3\right)$$

$$L[y] = \frac{1}{p^2(p^2-1)} + \frac{p}{p^2-1} - \frac{3}{p^2-1}$$

$$L[y] = \frac{1}{p^2(p^2-1)} + \frac{p}{p^2-1}\frac{p^2}{p^2} - \frac{3}{p^2-1}\frac{p^2}{p^2}$$

$$L[y] = \frac{1}{p^2(p^2-1)} + \frac{p^3}{p^2(p^2-1)} - \frac{3p^2}{p^2(p^2-1)}$$

$$L[y] = \frac{1 + p^3 - 3p^2}{p^2(p^2-1)}$$

$$L[y] = \frac{p^3 - 3p^2 + 1}{p^2(p^2-1)}$$

$$L^{-1}[L[y]] = L^{-1}\left[\frac{p^3 - 3p^2 + 1}{p^2(p^2-1)}\right]$$

$$y = L^{-1}\left[\frac{p^3 - 3p^2 + 1}{p^2(p^2-1)}\right]$$

X

(425)

Chapter 21 Second-Order Nonhomogeneous Initial Value Problems

$$\frac{\rho^3 - 3\rho^2 + 1}{\rho \cdot \rho \, (\rho-1)(\rho+1)} = \frac{A}{\rho} + \frac{B}{\rho^2} + \frac{C}{\rho-1} + \frac{D}{\rho+1}$$

$$= \frac{A}{\rho} \frac{\rho}{\rho} \frac{\rho-1}{\rho-1} \frac{\rho+1}{\rho+1}$$

$$+ \frac{B}{\rho^2} \frac{\rho-1}{\rho-1} \frac{\rho+1}{\rho+1}$$

$$+ \frac{C}{\rho-1} \frac{\rho^2}{\rho^2} \frac{\rho+1}{\rho+1}$$

$$+ \frac{D}{\rho+1} \frac{\rho^2}{\rho^2} \frac{\rho-1}{\rho-1}$$

$$= \frac{A \, \rho \, (\rho^2-1)}{\rho^2 \, (\rho^2-1)}$$

$$+ \frac{B \, (\rho^2-1)}{\rho^2 \, (\rho^2-1)}$$

$$+ \frac{C \, (\rho^3 + \rho^2)}{\rho^2 \, (\rho^2-1)}$$

$$+ \frac{D \, (\rho^3 - \rho^2)}{\rho^2 \, (\rho^2-1)}$$

Chapter 21 Second-Order Nonhomogeneous Initial value Problem

$$= \frac{A(p^3 - p)}{p^2(p^2-1)}$$

$$+ \frac{B(p^2-1)}{p^2(p^2-1)}$$

$$+ \frac{C(p^3 + p^2)}{p^2(p^2-1)}$$

$$+ \frac{D(p^3 - p^2)}{p^2(p^2-1)}$$

$$= \frac{A(p^3 - p) + B(p^2-1) + C(p^3 + p^2) + D(p^3 - p^2)}{p^3 - 3p^2 + 1}$$

$$= \frac{Ap^3 - Ap + Bp^2 - B + Cp^3 + Cp^2 + Dp^3 - Dp^2}{p^3 - 3p^2 + 1}$$

$$= \frac{Ap^3 + Cp^3 + Dp^3 + Bp^2 + Cp^2 - Dp^2 - Ap - B}{p^3 - 3p^2 + 1}$$

$$= \frac{(A + C + D)p^3 + (B + C - D)p^2 - Ap - B}{p^3 - 3p^2 + 1}$$

$A + C + D = 1$

$B + C - D = -3$

$-A = 0$

$-B = 1$

Chapt 21 Second-Order Nonhomogenous Initial Value Problem

$A = 0$, $B = -1$

$A + C + D = 1$

$0 + C + D = 1$

$C + D = 1$

$B + C - D = -3$

$-1 + C - D = -3$

$C - D = -3 + 1$

$C - D = -2$

$C + D = 1$

$C - D = -2$

$$C = \frac{\begin{vmatrix} 1 & 1 \\ -2 & -1 \end{vmatrix}}{\begin{vmatrix} 1 & 1 \\ 1 & -1 \end{vmatrix}} = \frac{-1 + 2}{-1 - 1} = \frac{1}{-2} = -\frac{1}{2}$$

$$D = \frac{\begin{vmatrix} 1 & 1 \\ 1 & -2 \end{vmatrix}}{\begin{vmatrix} 1 & 1 \\ 1 & -1 \end{vmatrix}} = \frac{-2 - 1}{-1 - 1} = \frac{-3}{-2} = \frac{3}{2}$$

$A = 0$, $B = -1$, $C = -\frac{1}{2}$, $D = \frac{3}{2}$

$$\frac{0}{p} - \frac{1}{p^2} - \frac{\frac{1}{2}}{p-1} + \frac{\frac{3}{2}}{p+1}$$

$$= -\frac{1}{p^2} - \frac{\frac{1}{2}}{p-1} + \frac{\frac{3}{2}}{p+1}$$

Chapter 21 Second-Order Nonhomogeneous Initial Value Problems

$$y = L^{-1}\left[\frac{p^3 - 3p^2 + 1}{p^2(p^2-1)}\right]$$

$$y = L^{-1}\left[-\frac{1}{p^2} - \frac{\frac{1}{2}}{p-1} + \frac{\frac{3}{2}}{p+1}\right]$$

$$y = L^{-1}\left[\frac{\frac{3}{2}}{p+1} - \frac{\frac{1}{2}}{p-1} - \frac{1}{p^2}\right]$$

$$y = L^{-1}\left[\frac{\frac{3}{2}}{p+1}\right] - L^{-1}\left[\frac{\frac{1}{2}}{p-1}\right] - L^{-1}\left[\frac{1}{p^2}\right]$$

$$y = \frac{3}{2}L^{-1}\left[\frac{1}{p+1}\right] - \frac{1}{2}L^{-1}\left[\frac{1}{p-1}\right] - L^{-1}\left[\frac{1!}{p^{1+1}}\right]$$

$$y = \frac{3}{2}e^{-x} - \frac{1}{2}e^{x} - x^{1}$$

$$y = \frac{3}{2}e^{-x} - \frac{1}{2}e^{x} - x$$

(429)

Chapt 21 Second-Order Nonhomogeneous Initial Value Problem
The Conventional Method Or
The Laplace Transform Method
For Solving I. V. P.'s

① which method should be used for solving
initial value problems —
ⓐ the conventional method   "
ⓑ the Laplace transform method
depends upon the personal preference
of the student, physicist or engineer
solving the I. V. P.

② the Laplace transform method has the advantage
of not having to solve the D.E.

③ it avoids that by dissolving the D.E.
into Laplace transforms

④ the inverse Laplace transform then recovers
the proper function that is the
correct solution to the I. V. P.

(430)

chapt 21 second-Order Nonhomogeneous I.V. P's

① solve the initial value problem
using two different methods
the conventional method   or
the Laplace transform method

$$\frac{d^2 y}{dx^2} - \frac{dy}{dx} - 2y = 5$$

$$y(0) = 2$$
$$y'(0) = -1$$

(cont)

Chapt 21 Second-Order Nonhomogeneous I.V.P's
the conventional method

$$\frac{d^2y}{dx^2} - \frac{dy}{dx} - 2y = 5$$

$$\frac{d^2y}{dx^2} - \frac{dy}{dx} - 2y = 0$$

$$m^2 - m - 2 = 0$$

$$(m-2)(m+1) = 0$$

$$m - 2 = 0 \qquad m + 1 = 0$$

$$m = 2 \qquad m = -1$$

$$m = 2, -1$$

$$y = c_1 e^{m_1 x} + c_2 e^{m_2 x}$$

$$y = c_1 e^{2x} + c_2 e^{-x}$$

$$y_h = c_1 e^{2x} + c_2 e^{-x}$$

$$y = y_h + y_p$$

Chapt 21 Second-Order Nonhomogeneous I.V.P.'s

1 cont

finding y_p

the nonhomogeneous term is 5

to family is

[1]

$y_p = A \cdot 1$

$y_p = A$

$y'_p = 0$

$y''_p = 0$

$y'' - y' - 2y = 5$

$0 - 0 - 2A = 5$

$-2A = 5$

$A = -\dfrac{5}{2}$

$y_p = A$

$y_p = -\dfrac{5}{2}$

(433)

(16nt)

Chpt 21 Second - Order Nonhomogeneous I.V.P:

$y = y_h + y_r$

$y = c_1 e^{2x} + c_2 e^{-x} - \frac{5}{2}$

Chapt 21 Second-Order Nonhomogeneous I. V. P's

$\underline{y(0) = 2}$ $\qquad x = 0, y = 2$

$y = c_1 e^{2x} + c_2 e^{-x} - \frac{5}{2}$

$2 = c_1 e^0 + c_2 e^0 - \frac{5}{2}$

$2 = c_1 (1) + c_2 (1) - \frac{5}{2}$

$2 = c_1 + c_2 - \frac{5}{2}$

$c_1 + c_2 = 2 + \frac{5}{2}$

$c_1 + c_2 = \frac{9}{2}$

$2c_1 + 2c_2 = 9$

$\underline{y'(0) = -1} \qquad x = 0, y' = -1$

$y = c_1 e^{2x} + c_2 e^{-x} - \frac{5}{2}$

$y' = c_1 e^{2x}(2) + c_2 e^{-x}(-1) - 0$

$y' = 2c_1 e^{2x} - c_2 e^{-x}$

$-1 = 2c_1 e^0 - c_2 e^0$

$-1 = 2c_1 (1) - c_2 (1)$

$-1 = 2c_1 - c_2$

$2c_1 - c_2 = -1$

$2c_1 + 2c_2 = 9$

$2c_1 - c_2 = -1$

Chapt 21 Second-Order Nonhomogeneous I.V.P's

$$2C_1 + 2C_2 = 9$$
$$2C_1 - C_2 = -1$$

$$C_1 = \frac{\begin{vmatrix} 9 & 2 \\ -1 & -1 \end{vmatrix}}{\begin{vmatrix} 2 & 2 \\ 2 & -1 \end{vmatrix}} = \frac{-9 + 2}{-2 - 4} = \frac{-7}{-6} = \frac{7}{6}$$

$$C_2 = \frac{\begin{vmatrix} 2 & 9 \\ 2 & -1 \end{vmatrix}}{\begin{vmatrix} 2 & 2 \\ 2 & -1 \end{vmatrix}} = \frac{-2 - 18}{-2 - 4} = \frac{-20}{-6} = \frac{10}{3}$$

$$y = C_1 e^{2x} + C_2 e^{-x} - \frac{5}{2}$$
$$y = \frac{7}{6} e^{2x} + \frac{10}{3} e^{-x} - \frac{5}{2}$$

$$\boxed{436}$$

Chapt 21 Second - Order Nonhomogeneous I.V.P.'s

Check

$$y'' - y' - 2y = 5$$

$$y = \tfrac{7}{6} e^{2x} + \tfrac{10}{3} e^{-x} - \tfrac{5}{2}$$
$$y' = \tfrac{7}{6} e^{2x}(2) + \tfrac{10}{3} e^{-x}(-1) - 0$$
$$y' = \tfrac{7}{3} e^{2x} - \tfrac{10}{3} e^{-x}$$
$$y'' = \tfrac{7}{3} e^{2x}(2) - \tfrac{10}{3} e^{-x}(-1)$$
$$y'' = \tfrac{14}{3} e^{2x} + \tfrac{10}{3} e^{-x}$$

$$\tfrac{14}{3} e^{2x} + \tfrac{10}{3} e^{-x}$$

$$- \left(\tfrac{7}{3} e^{2x} - \tfrac{10}{3} e^{-x} \right)$$

$$- 2 \left(\tfrac{7}{6} e^{2x} + \tfrac{10}{3} e^{-x} - \tfrac{5}{2} \right)$$

$$= 5$$

$$\tfrac{14}{3} e^{2x} + \tfrac{10}{3} e^{-x}$$

$$- \tfrac{7}{3} e^{2x} + \tfrac{10}{3} e^{-x}$$

$$- \tfrac{7}{3} e^{2x} - \tfrac{20}{3} e^{-x} + 5$$

$$= 5$$

$$5 = 5$$

(437)

Chapt 21 Second-Order Nonhomogeneous I.V.P.

(cont)

$y(0) = 2 \qquad\qquad X = 0, \; y = 2$

$y = \frac{7}{6} e^{2x} + \frac{10}{3} e^{-x} - \frac{5}{2}$

$y = \frac{7}{6} e^{0} + \frac{10}{3} e^{0} - \frac{5}{2}$

$y = \frac{7}{6}(1) + \frac{10}{3}(1) - \frac{5}{2}$

$y = \frac{7}{6} + \frac{10}{3} - \frac{5}{2}$

$y = \frac{7}{6} + \frac{20}{6} - \frac{15}{6}$

$y = \frac{12}{6}$

$y = 2$

$y'(0) = -1 \qquad\qquad X = 0, \; y' = -1$

$y = \frac{7}{6} e^{2x} + \frac{10}{3} e^{-x} - \frac{5}{2}$

$y' = \frac{7}{6} e^{2x} (2) + \frac{10}{3} e^{-x} (-1) - 0$

$y' = \frac{7}{3} e^{2x} - \frac{10}{3} e^{-x}$

$y' = \frac{7}{3} e^{0} - \frac{10}{3} e^{0}$

$y' = \frac{7}{3}(1) - \frac{10}{3}(1)$

$y' = \frac{7}{3} - \frac{10}{3}$

$y' = -\frac{3}{3}$

$y' = -1$

$\boxed{438}$

Chapt 21 Second-Order Nonhomogeneous I.V.P.'s using The Laplace transform method

$$\frac{d^2y}{dx^2} - \frac{dy}{dx} - 2y = 5$$

$$y(0) = 2$$
$$y'(0) = -1$$

$$\frac{d^2y}{dx^2} - \frac{dy}{dx} - 2y = 5$$

$$L\left[\frac{d^2y}{dx^2} - \frac{dy}{dx} - 2y\right] = L[5]$$

$$L\left[\frac{d^2y}{dx^2}\right] - L\left[\frac{dy}{dx}\right] - L[2y] = L[5]$$

$$L\left[\frac{d^2y}{dx^2}\right] - L\left[\frac{dy}{dx}\right] - 2L[y] = L[5]$$

$$L[y'] = p\,L[y] - y(0)$$
$$L[y''] = p^2\,L[y] - p\,y(0) - y'(0)$$

$$p^2 L[y] - p\,y(0) - y'(0) - \{p\,L[y] - y(0)\} - 2L[y] = L[5]$$

$$p^2 L[y] - p\,y(0) - y'(0) - p\,L[y] + y(0) - 2L[y] = L[5]$$

$$p^2 L[y] - p(2) + 1 - p\,L[y] + 2 - 2L[y] = L[5]$$

$$p^2 L[y] - 2p + 3 - p\,L[y] - 2L[y] = L[5]$$

$$p^2 L[y] - p\,L[y] - 2L[y] = L[5] + 2p - 3$$

$$L[y]\,(p^2 - p - 2) = \frac{5}{p} + 2p - 3$$

$$L[y] = \left(\frac{1}{p^2 - p - 2}\right)\left(\frac{5}{p} + 2p - 3\right)$$

(439)

(cont) Chpt 21 Second-Order Nonhomogeneous I.V.P.'s

$$L[y] = \left(\frac{1}{p^2 - p - 2}\right)\left(\frac{5}{p} + \frac{2p^2}{p} - \frac{3p}{p}\right)$$

$$L[y] = \left(\frac{1}{p^2 - p - 2}\right)\left(\frac{2p^2 - 3p + 5}{p}\right)$$

$$L[y] = \frac{2p^2 - 3p + 5}{p(p^2 - p - 2)}$$

$$L[y] = \frac{2p^2 - 3p + 5}{p(p - 2)(p + 1)}$$

$$L[y] = \frac{2p^2 - 3p + 5}{p(p + 1)(p - 2)}$$

$$L^{-1}[L[y]] = L^{-1}\left[\frac{2p^2 - 3p + 5}{p(p + 1) p - 2}\right]$$

$$y = L^{-1}\left[\frac{2p^2 - 3p + 5}{p(p + 1)(p - 2)}\right]$$

Chapt 21 Second-Order Non homogeneous I.V.P.'s

(1 cont) $\dfrac{2p^2 - 3p + 5}{p(p+1)(p-2)} = \dfrac{A}{p} + \dfrac{B}{p+1} + \dfrac{C}{p-2}$

$$= \dfrac{A}{p}\dfrac{p+1}{p+1}\dfrac{p-2}{p-2} + \dfrac{B}{p+1}\dfrac{p}{p}\dfrac{p-2}{p-2} + \dfrac{C}{p-2}\dfrac{p}{p}\dfrac{p+1}{p+1}$$

$$= \dfrac{A(p+1)(p-2) + B\,p\,(p-2) + C\,p\,(p+1)}{p(p+1)(p-2)}$$

$$= \dfrac{A(p^2 - p - 2) + B(p^2 - 2p) + C(p^2 + p)}{p(p+1)(p-2)}$$

$A(p^2 - p - 2) + B(p^2 - 2p) + C(p^2 + p) = 2p^2 - 3p + 5$

$Ap^2 - Ap - 2A + Bp^2 - 2Bp + Cp^2 + Cp = 2p^2 - 3p + 5$

$Ap^2 + Bp^2 + Cp^2 - Ap - 2Bp + Cp - 2A = 2p^2 - 3p + 5$

$(A + B + C)p^2 - (A + 2B - C)p - 2A = 2p^2 - 3p + 5$

$A + B + C = 2 \qquad -(A + 2B - C) = -3 \qquad -2A = 5$

$\qquad\qquad\qquad\qquad A + 2B - C = 3 \qquad\qquad A = -\dfrac{5}{2}$

$-\dfrac{5}{2} + B + C = 2 \qquad -\dfrac{5}{2} + 2B - C = 3$

$-5 + 2B + 2C = 4 \qquad -5 + 4B - 2C = 6$

$2B + 2C = 9 \qquad\qquad 4B - 2C = 11$

$2B + 2C = 9$

$4B - 2C = 11$

(441)

(1cnt) Chpt21 Second-Order Nonhomogeneous I.V.P's

$$2B + 2C = 9$$
$$4B - 2C = 11$$

$$B = \frac{\begin{vmatrix} 9 & 2 \\ 11 & -2 \end{vmatrix}}{\begin{vmatrix} 2 & 2 \\ 4 & -2 \end{vmatrix}} = \frac{-18-22}{-4-8} = \frac{-40}{-12}$$

$$= \frac{10}{3}$$

$$C = \frac{\begin{vmatrix} 2 & 9 \\ 4 & 11 \end{vmatrix}}{\begin{vmatrix} 2 & 2 \\ 4 & -2 \end{vmatrix}} = \frac{22-36}{-4-8} = \frac{-14}{-12} = \frac{7}{6}$$

$$A = -\frac{5}{2}, \quad B = \frac{10}{3}, \quad C = \frac{7}{6}$$

$$\frac{A}{p} + \frac{B}{p+1} + \frac{C}{p-2}$$

$$= \frac{-\frac{5}{2}}{p} + \frac{\frac{10}{3}}{p+1} + \frac{\frac{7}{6}}{p-2}$$

$$= \frac{\frac{7}{6}}{p-2} + \frac{\frac{10}{3}}{p+1} - \frac{\frac{5}{2}}{p}$$

Class # 21   Second-Order Nonhomogeneous I.V.P.:

(1 Cont)

$$y = L^{-1}\left[\frac{2p^2 - 3p + 5}{p(p+1)(p-2)}\right]$$

$$y = L^{-1}\left[\frac{\frac{7}{6}}{p-2} + \frac{\frac{10}{3}}{p+1} - \frac{\frac{5}{2}}{p}\right]$$

$$y = L^{-1}\left[\frac{\frac{7}{6}}{p-2}\right] + L^{-1}\left[\frac{\frac{10}{3}}{p+1}\right] - L^{-1}\left[\frac{\frac{5}{2}}{p}\right]$$

$$y = \frac{7}{6}L^{-1}\left[\frac{1}{p-2}\right] + \frac{10}{3}L^{-1}\left[\frac{1}{p+1}\right] - \frac{5}{2}L^{-1}\left[\frac{1}{p}\right]$$

$$y = \frac{7}{6}e^{2x} + \frac{10}{3}e^{-x} - \frac{5}{2} \quad (1)$$

$$y = \frac{7}{6}e^{2x} + \frac{10}{3}e^{-x} - \frac{5}{2}$$

(443)

Chapt 21 Second-Order Nonhomogeneous I.V.P's

2) solve the initial value problem
using two different methods
the Conventional method and
the Laplace transform method

$$\frac{d^2 y}{dx^2} - y = -3x$$

$$y(0) = -2$$
$$y'(0) = 4$$

444

Chapt 21 Second-Order Nonhomogeneous I.V.P.'s
the convential method

$$\frac{d^2y}{dx^2} - y = -3x$$

$$\frac{d^2y}{dx^2} - y = 0$$

$$m^2 - 1 = 0$$

$$m^2 = 1$$

$$m = \pm\sqrt{1}$$

$$m = \pm 1$$

$$m = 1, -1$$

$$y = c_1 e^{m_1 x} + c_2 e^{m_2 x}$$

$$y = c_1 e^x + c_2 e^{-x}$$

$$y_h = c_1 e^x + c_2 e^{-x}$$

$$y = y_h + y_p$$

Chapter 21 Second-Order Nonhomogeneous I.V.P's

6. Find y_p

the non homogeneous term is $-3X$

the family is

$\{X, 1\}$

$y_p = AX + B \cdot 1$
$y_p = AX + B$
$y'_p = A$
$y''_p = 0$

$y'' - y = -3X$
$0 - (AX + B) = -3X$
$-AX - B = -3X$
$AX + B = 3X$
$A = 3 \qquad B = 0$

$y_p = AX + B$
$y_p = 3X + 0$
$y_p = 3X$

$$\boxed{446}$$

Chapt 21 Second-Order Nonhomogeneous I.V.P's

$$y = y_h + y_p$$

$$y = c_1 e^x + c_2 e^{-x} + 3x$$

II E $\quad\quad\quad\quad\quad\quad\quad\quad\quad\quad\quad\quad\quad\quad\quad\quad\quad\quad$ P_{rob}

(447)

Chapt 21 Second-Order Nonhomogeneous I. V. P.'s

(2 cont)

$$\underline{y(0) = -2} \quad\quad\quad x = 0, \quad y = -2$$

$$y = c_1 e^x + c_2 e^{-x} + 3x$$

$$-2 = c_1 e^0 + c_2 e^0 + 0$$

$$-2 = c_1(1) + c_2(1)$$

$$-2 = c_1 + c_2$$

$$c_1 + c_2 = -2$$

$$\underline{y'(0) = 4} \quad\quad\quad x = 0, \quad y' = 4$$

$$y = c_1 e^x + c_2 e^{-x} + 3x$$

$$y' = c_1 e^x - c_2 e^{-x} + 3$$

$$4 = c_1 e^0 - c_2 e^0 + 3$$

$$4 = c_1(1) - c_2(1) + 3$$

$$4 = c_1 - c_2 + 3$$

$$c_1 - c_2 = 4 - 3$$

$$c_1 - c_2 = 1$$

$$c_1 + c_2 = -2$$

$$c_1 - c_2 = 1$$

$$\boxed{448}$$

Chapt 21 Second-Order Nonhomogeneous I.V.P's

$\boxed{2 \text{ cont}}$

$$c_1 + c_2 = -2$$
$$c_1 - c_2 = 1$$

$$C_1 = \frac{\begin{vmatrix} -2 & 1 \\ 1 & -1 \end{vmatrix}}{\begin{vmatrix} 1 & 1 \\ 1 & -1 \end{vmatrix}} = \frac{2-1}{-1-1} = \frac{1}{-2} = -\frac{1}{2}$$

$$C_2 = \frac{\begin{vmatrix} 1 & -2 \\ 1 & 1 \end{vmatrix}}{\begin{vmatrix} 1 & 1 \\ 1 & -1 \end{vmatrix}} = \frac{1+2}{-1-1} = \frac{3}{-2} = -\frac{3}{2}$$

$$y = C_1 e^x + C_2 e^{-x} + 3x$$
$$y = -\frac{1}{2} e^x - \frac{3}{2} e^{-x} + 3x$$

II G   Prob

2 cont | Chapt 21 Second-Order Nonhomogeneous I.V.P.

Check

$$y'' - y = -3x$$

$$y = -\frac{1}{2}e^x - \frac{3}{2}e^{-x} + 3x$$

$$y' = -\frac{1}{2}e^x + \frac{3}{2}e^{-x} + 3$$

$$y'' = -\frac{1}{2}e^x - \frac{3}{2}e^{-x}$$

$$\left(-\frac{1}{2}e^x - \frac{3}{2}e^{-x}\right) - \left(-\frac{1}{2}e^x - \frac{3}{2}e^{-x} + 3x\right) = -3x$$

$$-\frac{1}{2}e^x - \frac{3}{2}e^{-x} + \frac{1}{2}e^x + \frac{3}{2}e^{-x} - 3x = -3x$$

$$-3x = -3x$$

$$\underline{y(0) = -2} \qquad x = 0, \quad y = -2$$

$$y = -\frac{1}{2}e^x - \frac{3}{2}e^{-x} + 3x$$

$$y = -\frac{1}{2}e^0 - \frac{3}{2}e^0 + 0$$

$$y = -\frac{1}{2}(1) - \frac{3}{2}(1)$$

$$y = -\frac{1}{2} - \frac{3}{2}$$

$$y = -\frac{4}{2}$$

$$y = -2$$

$$\underline{y'(0) = 4} \qquad\qquad x = 0, \quad y' = 4$$

$$y = -\frac{1}{2}e^x - \frac{3}{2}e^{-x} + 3x$$

$$y' = -\frac{1}{2}e^x + \frac{3}{2}e^{-x} + 3$$

$$y' = -\frac{1}{2}e^0 + \frac{3}{2}e^0 + 3$$

$$y' = -\frac{1}{2}(1) + \frac{3}{2}(1) + 3$$

$$y' = -\frac{1}{2} + \frac{3}{2} + 3$$

$$y' = 4$$

(7 cont)

Chapt 21 Second-Order Nonhomogeneous I.V.P's

the Laplace transform method

$$\frac{d^2y}{dx^2} - y = -3x$$

$$y(0) = -2$$

$$y'(0) = 4$$

$$\frac{d^2y}{dx^2} - y = -3x$$

$$L\left[\frac{d^2y}{dx^2} - y\right] = L[-3x]$$

$$L\left[\frac{d^2y}{dx^2}\right] - L[y] = -3\,L[x]$$

$$L[y'] = p\,L[y] - y(0)$$

$$L[y''] = p^2\,L[y] - p\,y(0) - y'(0)$$

$$p^2\,L[y] - p\,y(0) - y'(0) - L[y] = -3\,L[x]$$

$$p^2\,L[y] - p(-2) - 4 - L[y] = -3\,L[x]$$

$$p^2\,L[y] + 2p - 4 - L[y] = -3\,L[x]$$

$$p^2\,L[y] - L[y] = -2p + 4 - 3\,L[x]$$

$$L[y]\,(p^2 - 1) = -2p + 4 - 3\left(\frac{1!}{p^{1+1}}\right)$$

$$L[y] = \frac{1}{p^2 - 1}\left(-2p + 4 - \frac{3}{p^2}\right)$$

$$L[y] = \frac{1}{p^2 - 1}\left(4 - 2p - \frac{3}{p^2}\right)$$

II √   Prob

(451)

Chapt 21 Second-Order Nonhomogeneous I.V.P.'s

(2 cont)
$$L[y] = \left(\frac{1}{p^2-1}\right)\left(\frac{4p^2}{p^2} - \frac{2p^3}{p^2} - \frac{3}{p^2}\right)$$

$$L[y] = \left(\frac{1}{p^2-1}\right)\left(\frac{4p^2 - 2p^3 - 3}{p^2}\right)$$

$$L[y] = \frac{4p^2 - 2p^3 - 3}{p^2(p^2-1)}$$

$$L[y] = \frac{4p^2 - 2p^3 - 3}{p^2(p-1)(p+1)}$$

$$L^{-1}[L[y]] = L^{-1}\left[\frac{4p^2 - 2p^3 - 3}{p^2(p+1)(p-1)}\right]$$

$$y = L^{-1}\left[\frac{-2p^3 + 4p^2 - 3}{p^2(p+1)(p-1)}\right]$$

Unit 21 Second-Order Nonhomogeneous I.V.P.'s

(Σ cont)
$$\frac{-2p^3 + 4p^2 - 3}{p^2(p+1)(p-1)}$$

$$= \frac{A}{p}$$

$$+ \frac{B}{p^2}$$

$$+ \frac{C}{p+1}$$

$$+ \frac{D}{p-1}$$

$$= \frac{A}{p} \cdot \frac{p}{p} \cdot \frac{p+1}{p+1} \cdot \frac{p-1}{p-1}$$

$$+ \frac{B}{p^2} \cdot \frac{p+1}{p+1} \cdot \frac{p-1}{p-1}$$

$$+ \frac{C}{p+1} \cdot \frac{p^2}{p^2} \cdot \frac{p-1}{p-1}$$

$$+ \frac{D}{p-1} \cdot \frac{p^2}{p^2} \cdot \frac{p+1}{p+1}$$

Chapt 21 Second-Order Nonhomogeneous I.V.P's

$$\left(2 \cos t\right) = \frac{A \quad p\left(p^2-1\right)}{p^2\left(p+1\right)\left(p-1\right)}$$

$$+ \frac{B\left(p^2-1\right)}{p^2\left(p+1\right)\left(p-1\right)}$$

$$+ \frac{C\left(p^3 - p^2\right)}{p^2\left(p+1\right)\left(p-1\right)}$$

$$+ \frac{D\left(p^3 + p^2\right)}{p^2\left(p+1\right)\left(p-1\right)}$$

$$= \frac{A\left(p^3 - p\right)}{p^2\left(p+1\right)\left(p-1\right)}$$

$$+ \frac{B\left(p^2-1\right)}{p^2\left(p+1\right)\left(p-1\right)}$$

$$+ \frac{C\left(p^3 - p^2\right)}{p^2\left(p+1\right)\left(p-1\right)}$$

$$+ \frac{D\left(p^3 + p^2\right)}{p^2\left(p+1\right)\left(p-1\right)}$$

(2 cont) $A(p^3 - p) + B(p^2 - 1) + C(p^3 - p^2) + D(p^3 + p^2)$
$= -2p^3 + 4p^2 - 3$

$Ap^3 - Ap + Bp^2 - B + Cp^3 - Cp^2 + Dp^3 + Dp^2$
$= -2p^3 + 4p^2 - 3$

$Ap^3 + Cp^3 + Dp^3$
$+ Bp^2 - Cp^2 + Dp^2$
$- Ap$
$- B$
$= -2p^3 + 4p^2 - 3$

$(A + C + D) p^3$
$+ (B - C + D) p^2$
$- Ap$
$- B$
$= -2p^3 + 4p^2 - 3$

$A + C + D = -2$
$B - C + D = 4$
$- A = 0$
$- B = -3$

$A = 0, B = 3$

$$\boxed{455}$$

Chapt 21 Second-Order Nonhomogeneous I.V.P's

2 cont

$0 + C + D = -2$ $B - C + D = 4$

$C + D = -2$ $3 - C + D = 4$

$-C + D = 1$

$C - D = -1$

$C + D = -2$

$C - D = -1$

$$C = \frac{\begin{vmatrix} -2 & 1 \\ -1 & -1 \end{vmatrix}}{\begin{vmatrix} 1 & 1 \\ 1 & -1 \end{vmatrix}} = \frac{2+1}{-1-1} = \frac{3}{-2} = -\frac{3}{2}$$

$$D = \frac{\begin{vmatrix} 1 & -2 \\ 1 & -1 \end{vmatrix}}{\begin{vmatrix} 1 & 1 \\ 1 & -1 \end{vmatrix}} = \frac{-1+2}{-1-1} = \frac{1}{-2} = -\frac{1}{2}$$

$A = 0, \quad B = 3, \quad C = -\frac{3}{2}, \quad D = -\frac{1}{2}$

$$\frac{A}{p} + \frac{B}{p^2} + \frac{C}{p+1} + \frac{D}{p-1}$$

$$= \frac{0}{p} + \frac{3}{p^2} + \frac{-\frac{3}{2}}{p+1} + \frac{-\frac{1}{2}}{p-1}$$

$$= \frac{3}{p^2} - \frac{\frac{3}{2}}{p+1} - \frac{\frac{1}{2}}{p-1}$$

(456)

Chapt 21 Second-Order Nonhomogeneous I.V.P's

(2 cont) $y = L^{-1} \left[\dfrac{-2p^3 + 4p^2 - 3}{p^2 (p+1)(p-1)} \right]$

$y = L^{-1} \left[\dfrac{3}{p^2} - \dfrac{\frac{3}{2}}{p+1} - \dfrac{\frac{1}{2}}{p-1} \right]$

$y = L^{-1} \left[\dfrac{3}{p^2} \right] - L^{-1} \left[\dfrac{\frac{3}{2}}{p+1} \right] - L^{-1} \left[\dfrac{\frac{1}{2}}{p-1} \right]$

$y = 3 L^{-1} \left[\dfrac{1}{p^2} \right] - \dfrac{3}{2} L^{-1} \left[\dfrac{1}{p+1} \right] - \dfrac{1}{2} L^{-1} \left[\dfrac{1}{p-1} \right]$

$y = 3 L^{-1} \left[\dfrac{1!}{p^{1+1}} \right] - \dfrac{3}{2} L^{-1} \left[\dfrac{1}{p+1} \right] - \dfrac{1}{2} L^{-1} \left[\dfrac{1}{p-1} \right]$

$y = 3x - \dfrac{3}{2} e^{-x} - \dfrac{1}{2} e^{x}$

$y = -\dfrac{1}{2} e^{x} - \dfrac{3}{2} e^{-x} + 3x$

Chpt 21 Second-Order Nonhomogeneous I.V.P's

③ solve the initial value problem
using two different methods
the conventional method and
the Laplace transform method

$$\frac{d^2 y}{dx^2} + y = e^{-x}$$

$$y(0) = 5$$

$$y'(0) = -2$$

(458)

<u>Chapt 21 Second-Order Nonhomogeneous I.V.P's</u>

(3 cont) <u>the conventional method</u>

$$\frac{d^2 y}{dx^2} + y = e^{-x}$$

$$\frac{d^2 y}{dx^2} + y = 0$$

$$m^2 + 1 = 0$$
$$m^2 = -1$$
$$m = \pm\sqrt{-1}$$
$$m = \pm i$$

$$0 \pm i$$
$$a \pm b i$$
$$a = 0 \quad b = 1$$

$$y = e^{ax}(c_1 \cos b x + c_2 \sin b x)$$
$$y = e^{0}(c_1 \cos x + c_2 \sin x)$$
$$y = 1 (c_1 \cos x + c_2 \sin x)$$
$$y = c_1 \cos x + c_2 \sin x$$

$$y = y_h + y_p$$

(3 cont)

Chapt 21 Second-Order Nonhomogeneous I.V.P's

finding y_p

the nonhomogeneous term is e^{-x}

its family is

$\{e^{-x}\}$

$y_p = A e^{-x}$

$y'_p = -A e^{-x}$

$y''_p = A e^{-x}$

$y'' + y = e^{-x}$

$A e^{-x} + A e^{-x} = e^{-x}$

$2A e^{-x} = e^{-x}$

$2A = 1$

$A = \dfrac{1}{2}$

$y_p = A e^{-x}$

$y_p = \dfrac{1}{2} e^{-x}$

Chapt 21 Second-Order Nonhomogeneous I.V.P.'s

(3 cont) $y = y_h + y_p$

$y = c_1 \cos x + c_2 \sin x + \frac{1}{2} e^{-x}$

Chapt 21 Second-Order Nonhomogeneous I.V.P.'s

(3 cont) $\underline{y(0) = 5}$ $\qquad$ $x = 0, \ y = 5$

$y = C_1 \cos x + C_2 \sin x + \frac{1}{2} e^{-x}$

$5 = C_1 \cos 0 + C_2 \sin 0 + \frac{1}{2} e^{0}$

$5 = C_1 (1) + C_2 (0) + \frac{1}{2} (1)$

$5 = C_1 + \frac{1}{2}$

$C_1 = 5 - \frac{1}{2}$

$C_1 = \frac{10}{2} - \frac{1}{2}$

$C_1 = \frac{9}{2}$

$\underline{y'(0) = -2}$ $\qquad$ $x = 0, \ y' = -2$

$y = C_1 \cos x + C_2 \sin x + \frac{1}{2} e^{-x}$

$y' = -C_1 \sin x + C_2 \cos x - \frac{1}{2} e^{-x}$

$-2 = -C_1 \sin 0 + C_2 \cos 0 - \frac{1}{2} e^{0}$

$-2 = -C_1 (0) + C_2 (1) - \frac{1}{2} (1)$

$-2 = C_2 - \frac{1}{2}$

$C_2 = -2 + \frac{1}{2}$

$C_2 = -\frac{4}{2} + \frac{1}{2}$

$C_2 = -\frac{3}{2}$

$C_1 = \frac{9}{2}, \quad C_2 = -\frac{3}{2}$

$y = C_1 \cos x + C_2 \sin x + \frac{1}{2} e^{-x}$

$y = \frac{9}{2} \cos x - \frac{3}{2} \sin x + \frac{1}{2} e^{-x}$

(462)

Chapt 21 Second-Order Nonhomogeneous I.V.P's

check

$$y'' + y = e^{-x}$$

$$y = \frac{9}{2}\cos x - \frac{3}{2}\sin x + \frac{1}{2}e^{-x}$$

$$y' = -\frac{9}{2}\sin x - \frac{3}{2}\cos x - \frac{1}{2}e^{-x}$$

$$y'' = -\frac{9}{2}\cos x + \frac{3}{2}\sin x + \frac{1}{2}e^{-x}$$

$$-\frac{9}{2}\cos x + \frac{3}{2}\sin x + \frac{1}{2}e^{-x}$$

$$+ \quad \frac{9}{2}\cos x - \frac{3}{2}\sin x + \frac{1}{2}e^{-x}$$

$$= \quad e^{-x}$$

$$\frac{1}{2}e^{-x} + \frac{1}{2}e^{-x} = e^{-x}$$

$$e^{-x} = e^{-x}$$

$$\underline{y(0) = 5} \qquad x = 0, \; y = 5$$

$$y = \frac{9}{2}\cos x - \frac{3}{2}\sin x + \frac{1}{2}e^{-x}$$

$$y = \frac{9}{2}\cos 0 - \frac{3}{2}\sin 0 + \frac{1}{2}e^{0}$$

$$y = \frac{9}{2}(1) - \frac{3}{2}(0) + \frac{1}{2}(1)$$

$$y = \frac{9}{2} + \frac{1}{2} = \frac{10}{2} = 5$$

$$\underline{y'(0) = -2} \qquad x = 0, \; y' = -2$$

$$y = \frac{9}{2}\cos x - \frac{3}{2}\sin x + \frac{1}{2}e^{-x}$$

$$y' = -\frac{9}{2}\sin x - \frac{3}{2}\cos x - \frac{1}{2}e^{-x}$$

$$y' = -\frac{9}{2}\sin 0 - \frac{3}{2}\cos 0 - \frac{1}{2}e^{0}$$

$$y' = 0 - \frac{3}{2} - \frac{1}{2}$$

$$y' = -\frac{4}{2} = -2$$

Chapter 21 Second-Order Nonhomogeneous I.V.P.'s
the Laplace transform method

$$\frac{d^2y}{dx^2} + y = e^{-x}$$

$$y(0) = 5$$

$$y'(0) = -2$$

$$\frac{d^2y}{dx^2} + y = e^{-x}$$

$$L\left[\frac{d^2y}{dx^2} + y\right] = L[e^{-x}]$$

$$L\left[\frac{d^2y}{dx^2}\right] + L[y] = L[e^{-x}]$$

$$L[y'] = p\, L[y] - y(0)$$

$$L[y''] = p^2\, L[y] - p\, y(0) - y'(0)$$

$$p^2\, L[y] - p\, y(0) - y'(0) + L[y] = L[e^{-x}]$$

$$p^2\, L[y] - p(5) + 2 + L[y] = L[e^{-x}]$$

$$p^2\, L[y] - 5p + 2 + L[y] = L[e^{-x}]$$

$$p^2\, L[y] + L[y] = L[e^{-x}] + 5p - 2$$

$$L[y]\,(p^2+1) = \frac{1}{p+1} + 5p - 2$$

$$L[y] = \left(\frac{1}{p^2+1}\right)\left(\frac{1}{p+1} + 5p - 2\right)$$

$$L[y] = \frac{1}{(p^2+1)(p+1)} + \frac{5p}{p^2+1} - \frac{2}{p^2+1}$$

(4.64)

Chapt 21 Second-Order Nonhomogeneous I.V.P.'s

(3 cont) $L[y] = \dfrac{1}{(\rho^2+1)(\rho+1)} + \dfrac{5\rho}{\rho^2+1} \cdot \dfrac{\rho+1}{\rho+1} - \dfrac{2}{\rho^2+1} \cdot \dfrac{\rho+1}{\rho+1}$

$L[y] = \dfrac{1 + 5\rho(\rho+1) - 2(\rho+1)}{(\rho^2+1)(\rho+1)}$

$L[y] = \dfrac{1 + 5\rho^2 + 5\rho - 2\rho - 2}{(\rho+1)(\rho^2+1)}$

$L[y] = \dfrac{5\rho^2 + 3\rho - 1}{(\rho+1)(\rho^2+1)}$

$L^{-1}[L[y]] = L^{-1}\left[\dfrac{5\rho^2 + 3\rho - 1}{(\rho+1)(\rho^2+1)}\right]$

$y = L^{-1}\left[\dfrac{5\rho^2 + 3\rho - 1}{(\rho+1)(\rho^2+1)}\right]$

(465)

Chpt 21 Second-Order Nonhomogeneous I.V.P.:

(3 cont)

$$\frac{5p^2 + 3p - 1}{(p+1)(p^2+1)} = \frac{A}{p+1} + \frac{Bp+C}{p^2+1}$$

$$= \frac{A}{p+1}\frac{p^2+1}{p^2+1} + \frac{Bp+C}{p^2+1}\frac{p+1}{p+1}$$

$$= \frac{A(p^2+1) + (Bp+C)(p+1)}{(p+1)(p^2+1)}$$

$A(p^2+1) + (Bp+C)(p+1) = 5p^2 + 3p - 1$

$Ap^2 + A + Bp^2 + Bp + Cp + C = 5p^2 + 3p - 1$

$Ap^2 + Bp^2 + Bp + Cp + A + C = 5p^2 + 3p - 1$

$(A+B)p^2 + (B+C)p + (A+C) = 5p^2 + 3p - 1$

$A + B = 5 \qquad B + C = 3 \qquad A + C = -1$

$\qquad\qquad\qquad\qquad\qquad\qquad\qquad C = -1 - A$

$\qquad\qquad\qquad B - 1 - A = 3$

$\qquad\qquad\qquad B - A = 4$

$\qquad\qquad\qquad -A + B = 4$

$\qquad\qquad\qquad A - B = -4$

$A + B = 5$

$A - B = -4$

Chapter 21 Second-Order Nonhomogeneous I.V.P.

$A + B = 5$

$A - B = -4$

$$A = \frac{\begin{vmatrix} 5 & 1 \\ -4 & -1 \end{vmatrix}}{\begin{vmatrix} 1 & 1 \\ 1 & -1 \end{vmatrix}} = \frac{-5 + 4}{-1 - 1} = \frac{-1}{-2} = \frac{1}{2}$$

$$B = \frac{\begin{vmatrix} 1 & 5 \\ 1 & -4 \end{vmatrix}}{\begin{vmatrix} 1 & 1 \\ 1 & -1 \end{vmatrix}} = \frac{-4 - 5}{-1 - 1} = \frac{-9}{-2} = \frac{9}{2}$$

$B + C = 3$

$\frac{9}{2} + C = 3$

$C = 3 - \frac{9}{2}$

$C = \frac{6}{2} - \frac{9}{2}$

$C = -\frac{3}{2}$

$A = \frac{1}{2}, \quad B = \frac{9}{2}, \quad C = -\frac{3}{2}$

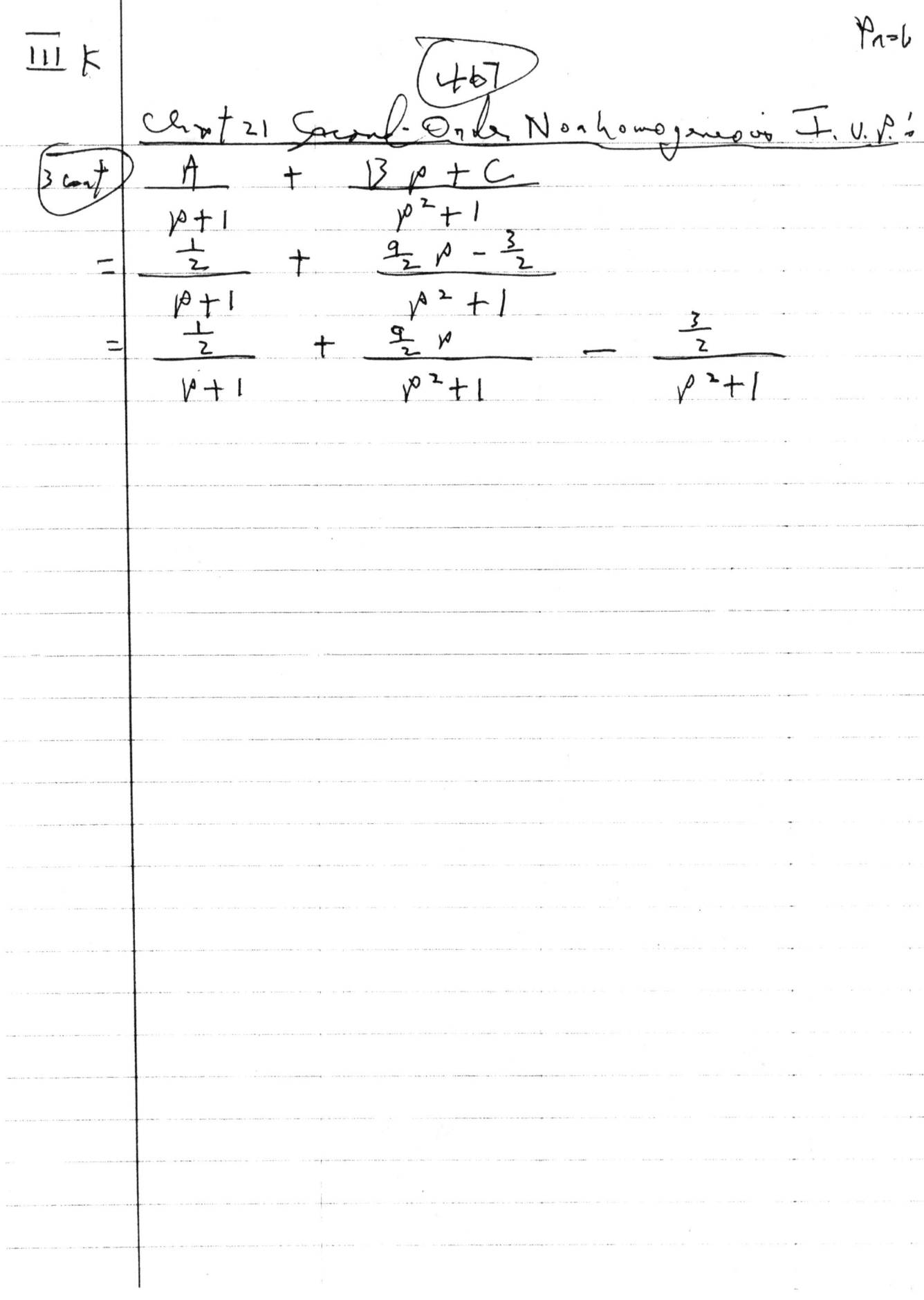

467

Chapt 21 Second-Order Nonhomogeneous I.V.P's

$\boxed{3\,\text{cont}}$

$$\frac{A}{p+1} + \frac{B\,p+C}{p^2+1}$$

$$= \frac{\frac{1}{2}}{p+1} + \frac{\frac{9}{2}\,p - \frac{3}{2}}{p^2+1}$$

$$= \frac{\frac{1}{2}}{p+1} + \frac{\frac{9}{2}\,p}{p^2+1} - \frac{\frac{3}{2}}{p^2+1}$$

(468.)

Chapt 21 Second - Order Nonhomogeneous I. V. P.'s

(3 cont)

$$y = L^{-1}\left[\frac{5p^2 + 3p - 1}{(p+1)(p^2+1)}\right]$$

$$y = L^{-1}\left[\frac{\frac{1}{2}}{p+1} + \frac{\frac{9}{2}p - \frac{3}{2}}{p^2+1}\right]$$

$$y = L^{-1}\left[\frac{\frac{1}{2}}{p+1} + \frac{\frac{9}{2}p}{p^2+1} - \frac{\frac{3}{2}}{p^2+1}\right]$$

$$y = L^{-1}\left[\frac{\frac{1}{2}}{p+1}\right] + L^{-1}\left[\frac{\frac{9}{2}p}{p^2+1}\right] - L^{-1}\left[\frac{\frac{3}{2}}{p^2+1}\right]$$

$$y = \frac{1}{2}L^{-1}\left[\frac{1}{p+1}\right] + \frac{9}{2}L^{-1}\left[\frac{p}{p^2+1}\right] - \frac{3}{2}L^{-1}\left[\frac{1}{p^2+1}\right]$$

$$y = \frac{1}{2}L^{-1}\left[\frac{1}{p+1}\right] + \frac{9}{2}L^{-1}\left[\frac{p}{p^2+1^2}\right] - \frac{3}{2}L^{-1}\left[\frac{1}{p^2+1^2}\right]$$

$$y = \frac{1}{2}e^{-x} + \frac{9}{2}\cos x - \frac{3}{2}\sin x$$

$$y = \frac{9}{2}\cos x - \frac{3}{2}\sin x + \frac{1}{2}e^{-x}$$

www.ingramcontent.com/pod-product-compliance
Lightning Source LLC
Chambersburg PA
CBHW081102170526
45165CB00008B/2302